Logic in Asia: Studia Logica Library

Logic in Asia: Studia Logica Library

This book series promotes the advance of scientific research within the field of logic in Asian countries. It strengthens the collaboration between researchers based in Asia with researchers across the international scientific community and offers a platform for presenting the results of their collaborations. One of the most prominent features of contemporary logic is its interdisciplinary character, combining mathematics, philosophy, modern computer science, and even the cognitive and social sciences. The aim of this book series is to provide a forum for current logic research, reflecting this trend in the field's development.

The series accepts books on any topic concerning logic in the broadest sense, i.e., books on contemporary formal logic, its applications and its relations to other disciplines. It accepts monographs and thematically coherent volumes addressing important developments in logic and presenting significant contributions to logical research. In addition, research works on the history of logical ideas, especially on the traditions in China and India, are welcome contributions.

The scope of the book series includes but is not limited to the following:

- Monographs written by researchers in Asian countries.
- Proceedings of conferences held in Asia, or edited by Asian researchers.
- Anthologies edited by researchers in Asia.
- Research works by scholars from other regions of the world, which fit the goal of "Logic in Asia".

The series discourages the submission of manuscripts that contain reprints of previously published material and/or manuscripts that are less than 165 pages/ 90,000 words in length.

Please also visit our webpage: http://tsinghualogic.net/logic-in-asia/background/

Relation with Studia Logica Library

This series is part of the Studia Logica Library, and is also connected to the journal Studia Logica. This connection does not imply any dependence on the Editorial Office of Studia Logica in terms of editorial operations, though the series maintains cooperative ties to the journal.

This book series is also a sister series to Trends in Logic and Outstanding Contributions to Logic.

For inquiries and to submit proposals, authors can contact the editors-in-chief Fenrong Liu at fenrong@tsinghua.edu.cn or Hiroakira Ono at ono@jaist.ac.jp.

More information about this series at http://www.springer.com/series/13080

Beishui Liao · Yì N. Wáng
Editors

Context, Conflict and Reasoning

Proceedings of the Fifth Asian Workshop on Philosophical Logic

Springer

Editors
Beishui Liao
Zhejiang University
Hangzhou, China

Yì N. Wáng
Zhejiang University
Hangzhou, China

ISSN 2364-4613 ISSN 2364-4621 (electronic)
Logic in Asia: Studia Logica Library
ISBN 978-981-15-7133-6 ISBN 978-981-15-7134-3 (eBook)
https://doi.org/10.1007/978-981-15-7134-3

This Springer imprint is published by the registered company Springer Nature Singapore Pte Ltd.
The registered company address is: 152 Beach Road, #21-01/04 Gateway East, Singapore 189721, Singapore

Preface

We hereby present the proceedings of the Fifth Asian Workshop on Philosophical Logic (AWPL 2020) that was planned to take place in Hangzhou, China, 7-9 April 2020. The AWPL workshops are devoted to promote communication and collaboration between researchers, within and outside Asia, in philosophical logics and related fields, with an emphasis on the interplay of formal methods and their applications. This is a fruitful field with a long history, yet still generating sparkling new ideas and results each year by the best and very active researchers from all around the world.

Several rounds of open calls for papers were sent out, inviting researchers to submit papers for their work in the areas of non-classical logics, philosophical logics, algebraic logics, and their applications in computer science, cognitive science, and social sciences. Twenty two submissions were received, with authors from countries in Asia, Europe and South America, and were then peer-reviewed by the international program committee. Among them eight long papers and four short papers were finally included in the proceedings, after careful revision based on the suggestions and comments from the reviewers. The topics of the accepted papers diverge, from the core of philosophical logics such as conditionals, epistemic and deontic logics, to the borders such as game models for linguistic studies, yet they spark new ideas under the title "Context, Conflict and Reasoning".

First, related to conflict and reasoning, Marcos Cramer introduces a paracomplete truth theory with a definable hierarchy of determinateness operators, such that one's rejection of the Liar sentence within the object language of his theory can be explained; Ming Hsiung studies paradoxes behind the Solovay sentences, by using the truth predicate instead of the provability predicate in the formalisation of the Solovay function; Ziyue Hu proposes a deontic logic of multiple roles based on the role-related deontic syllogism, capable of tolerating normative conflicts.

Second, about context and reasoning, Mengyuan Zhao, Zhong Yin, and Ziming Lu propose a game-theoretic model of ambiguous pronoun resolution, where the pronoun reference is not clearly determined in the context; Zhaoqing Xu studies the factivity problem of epistemic contextualism by using the formal language and formal models of epistemic logic; Junli Jiang investigates how faith diffuses in a social network that the same person may belong to different cultural circles and with different threshold values to their influence, by using dynamic epistemic logic; Jialiang Yan and Fenrong Liu introduce a first-order deontic event model to study those natural language sentences that contain both quantiers and deontic modalities, such that the failed monotonic inferences can be explained.

Third, the remaining part of this volume studies various aspects of philosophical logic. Lingyuan Ye proposes a relational approach to semantics theory, as an alternative to the existing model-theoretic approach and proof-theoretic approach. This new approach is based on the idea that the meaning of an expression within a language is fundamentally involved with the relations between that expression and other expressions in the very same language. Izumi Takeuti and Katsuhiko Sano study Kripke semantics based on graphs, where the semantics interprets a formula into a set of nodes in a graph. They propose a syntax of extended modal logic in which planarity is definable. Takahiro Sawasaki and Katsuhiko Sano present Hilbert-style systems and sequent calculi for some weaker versions of common sense modal predicate calculus. Eric Raidl develops a technique to generate logics for those conditionals strengthened by additional conditions, by transfering completeness results of a known conditional to a definable conditional. Yiyan Wang introduces an approach to explaining collective intentionality, to deal with the problem that the individualistic explanation cannot sufficiently explain the concept of collective intentionality.

We would like to thank the authors who submitted papers for their hard work on the frontiers of philosophical logics and their applications, and the program committee (Thomas Ågotnes, Nick Bezhanishvili, Marcos Cramer, Huimin Dong, Sujata Ghosh, Fengkui Ju, Hanti Lin, Fenrong Liu, Hu Liu, Xinwen Liu, Zachiri Mckenzie, Meiyun Guo,Hiroakira Ono, Eric Pacuit, Xavier Parent, R Ramanujam, Ji Ruan, Katsuhiko Sano, Chenwei Shi, Dag Westerståhl, Emil Weydert, Junhua Yu, Yan Zhang, and Shengyang Zhong) for their selecting the papers that make these wonderful proceedings. Meanwhile, we are grateful to Fenrong Liu and Hiroakira Ono, the editors in chief of this book series "Logic in Asia" (LIAA), for their supportive recommendation of this volume to LIAA, and to Fiona Wu and Leana Li, for their support in the process of publication of this volume.

Finally, we acknowledge that AWPL 2020 is financially supported by the Convergence Research Project for Brain Research and Artificial Intelligence, Zhejiang University.

Beishui Liao & Yì N. Wáng
Department of Philosophy
Zhejiang University
Hangzhou, April 2020

Table of Contents

Paracomplete truth theory with a definable hierarchy of determinateness operators*

Marcos Cramer

International Center for Computational Logic, TU Dresden, Germany

Abstract. One way to deal with the Liar paradox is the paracomplete approach to theories of truth that gives up proofs by contradiction and the Law of the Excluded Middle. This allows one to reject both the Liar sentence and its negation. The simplest paracomplete theory of truth is *KFS* due to Saul Kripke. At face value, this theory suffers from the problem that it cannot say anything about the Liar paradox, so a defender of this theory cannot explain their rejection of the Liar sentence within the language of *KFS*. This was one of the motivations for Hartry Field to extend *KFS* with a conditional that is not definable within *KFS*. With the help of this conditional, Field defines a determinateness operator that can be used to explain one's rejection of the Liar sentence within the object language of his theory. The determinateness operator can be transfinitely iterated to create stronger notions of determinateness required to explain the rejection of paradoxical sentences involving the determinateness operator. In this paper, we show that Field's complex extension of *KFS* is not required in order to express rejection of paradoxical sentences like the Liar sentence. Instead one can work with a transfinite hierarchy of determinateness operators that are definable in *KFS*. After defining this hierarchy of determinateness operators, we compare their properties with the transfinitely iterable determinateness operator due to Field.

1 Introduction

In everyday conversations, in scientific texts and in philosophical discussions we often make use of the predicate "true". So to get a better understanding of how we communicate our ideas and how we judge the correctness of our arguments, it is desirable to have a reasonable *theory of truth*, i.e. a logical formalism that contains the predicate *True*, that captures the inferences involving this predicate that we would intuitively deem acceptable, and that satisfies further rationality criteria like consistency.

In any formalism that contains the predicate *True* and captures come basic arithmetical reasoning, one can construct a *Liar sentence*, i.e. a sentence that asserts of itself that it is not true. If we apply some classically valid inferences combined with some intuitive inferences for the predicate *True* to such a Liar sentence, we can derive an inconsistency, and thus by some further classically

* This chapter is in its final form and it is not submitted to publication anywhere else.

B. Liao and Y. N. Wáng (eds.), *Context, Conflict and Reasoning*,
Logic in Asia: Studia Logica Library,
https://doi.org/10.1007/978-981-15-7134-3_1

valid inferences, we can derive every sentence of the language, rendering the formalism practically useless. This is called the *Liar paradox*. Any reasonable theory of truth needs to handle the Liar paradox in some way by putting some restrictions either on the intuitive inferences for the predicate *True* or on the rules of classical logic. Field [3] proposes a so-called *paracomplete* theory of truth that deals with the Liar paradox by restricting classical logic to the strong Kleene logic K_3, in which proof by contradictions are not unrestrictedly admissible and the Law of the Excluded Middle ($\varphi \vee \neg\varphi$ for any formula φ) does not hold unrestrictedly, but which differs from intuitionistic logic by still admitting double negation elimination and De Morgan's Laws.

This paracomplete approach can be traced back to the work of Kripke [4]. The formal theory of truth that is based on the semantic construction due to Kripke is usually called *KFS* and has K_3 as its underlying logic.

The stance towards the Liar paradox that a paracomplete theory of truth defends is that of rejecting both the Liar paradox and its negation. It seems desirable that this stance should be expressible and justifiable within the formal theory that the paracompletist puts forward. At face value, it seems that this cannot be achieved withing *KFS*. To overcome this limitation of *KFS*, Field [3] has introduced a determinateness operator, which allows one to say that the Liar paradox is not determinately true as a justification for rejecting the Liar sentence. The determinateness operator can be transfinitely iterated to create stronger notions of determinateness required to explain the rejection of paradoxical sentences involving the determinateness operator.

While Field's theory of truth is based on *KFS*, his determinateness operator is based on a conditional that is not definable within *KFS* but needs to be added to the theory. Field provides a complex semantic characterization of this conditional, but does not provide a proof-theoretic account of it.

In this paper we show that Field's complex extension of *KFS* is not required in order to express reasons for rejecting paradoxical sentences like the Liar sentence. Instead one can work with a transfinite hierarchy of determinateness operators that are definable in *KFS*. The definition of these determinateness operators is inspired by the well-founded semantics of logic programming [9,1]. We will define this hierarchy of determinateness operators, compare its properties to the transfinitely iterable determinateness operator due to Field, and briefly sketch how the construction of the determinateness operators can be modified to construct a conditional that satisfies some desirable properties.

2 The Liar Paradox

A Liar sentence is a sentence that asserts of itself that it is not true. An informal example of a Liar sentence is the following sentence that uses the determiner *this* to refer to itself:

This sentence is not true.

Given that the semantics of the word *this* depends a lot on the communicative context, logicians often prefer to work with more formal Liar sentences whose interpretation is completely independent of the communicative context. For this purpose, one usually works in a formal language that extends the standard first-order language of arithmetic $\mathcal{L}^{\text{arithm}}$ with a truth predicate *True*, yielding the extended first-order language $\mathcal{L}^{\text{arithm}}_{\mathit{True}}$. It is well-known that assuming some basic formal theory of arithmetic, e.g. Peano Arithmetic, one can define a Gödel numbering of any given formal language, i.e. an encoding of the syntax of that language in terms of natural numbers which maps each formula φ of the formal language to a unique natural number $\langle\varphi\rangle$ that can be used to talk about φ in the formal language. Intuitively, the intended meaning of $\mathit{True}(n)$ is that there exists a sentence φ of $\mathcal{L}^{\text{arithm}}_{\mathit{True}}$ such that $\langle\varphi\rangle = n$ and φ is true.

It is well-known that using a diagonalization technique due to Carnap, Gödel and Tarski one can construct a sentence $L \in \mathcal{L}^{\text{arithm}}_{\mathit{True}}$ for which one can prove in Peano Arithmetic (and indeed in various weaker theories of arithmetic) that

$$L \leftrightarrow \neg \mathit{True}(\langle L\rangle) \tag{1}$$

Given our intuitive interpretation of *True*, the sentence L is thus provably equivalent to the statement that L is not true. In other words, L is a Liar sentence, and unlike the informal Liar sentence presented above, it is a purely formal Liar sentence that does not depend on the interpretation of a context-dependent word like *this*.

Once we have constructed L, it seems like we can derive both $\neg L$ and L using standard rules of inference. We start with a proof by contradiction that establishes $\neg L$: Assume for a contradiction that L holds. In that case L is true, i.e. $\mathit{True}(\langle L\rangle)$ holds. But from (1) we get $\mathit{True}(\langle L\rangle) \rightarrow \neg L$, so by modus ponens we get $\neg L$. This contradicts our assumption that L is true. This completes the proof by contradiction, i.e. we can retract the assumption and deduce $\neg L$. But from (1) we have $\neg L \rightarrow \mathit{True}(\langle L\rangle)$, so by modus ponens we get $\mathit{True}(\langle L\rangle)$, i.e. we get that L is true. So we have deduced both $\neg L$ and L, a contradiction.

This is what is commonly called the *Liar paradox*. If we try to formalize this apparent proof in the proof calculus of natural deduction, we see that apart from the explicitly stated rules *modus ponens* (also called ($\rightarrow$-Elim)) and *proof by contradiction* (also called ($\neg$-Introd)), we implicitly made use of two further rules of inference that involve the predicate *True*:

(T-Introd)	$\varphi \models \mathit{True}(\langle\varphi\rangle)$
(T-Elim)	$\mathit{True}(\langle\varphi\rangle) \models \varphi$

These two rules are very compelling, because they seem to precisely characterize our intuitions about the meaning of the predicate *True*. What the Liar paradox shows is that these rules cannot be consistently combined with classical logic. Multiple avenues have been explored to deal with this:

– Those who want to keep classical logic fully in place need to reject at least one of (T-Introd) and (T-Elim). The most well-known theory of truth that works with classical logic is the so-called Kripke-Feferman theory KF, which accepts (T-Elim) but rejects (T-Introd) [2]. However, this theory has the awkward property of declaring "L, but L is not true".

- One can bite the bullet and accept that both L and $\neg L$ can be derived, but restrict classical logic so that this inconsistency does not lead to explosion, i.e. to the derivability of all sentences. This approach is called the *paraconsistent* approach and was first proposed by Priest [5].
- One can restrict the structural rules of inference that were left implicit in the above piece of informal reasoning and that allow one to use already derived sentences as premises for further derivations, as well as to use an assumption more than once in a derivation [6].
- One can give up one of those rules of inference that were explicitly used in the above elicitation of the Liar paradox. The most common rule of inference to be dropped is proof by contradiction ($\neg$-Introd). Dropping this rule is called the *paracomplete* approach, and it has recently gained traction due to Field's [3] defense of it.

In this paper we are working with a paracomplete approach to the Liar paradox, i.e. we are giving up proof by contradiction in its unrestricted form, which allows us to accept (T-Introd) and (T-Elim) unrestrictedly.

3 From Kripke to Field: Paracomplete Approaches to Semantic Paradoxes

Kripke [4] defined a construction that can be used to give a three-valued model-theoretic semantics for the language $\mathcal{L}^{\text{arithm}}_{True}$. This construction gives rise to the paracomplete theory *KFS* and is also at the heart of the paracomplete theory of truth presented by Field [3]. The same construction can also be used as the basis for approaches based on classical logic, e.g. for the Kripke-Feferman theory KF. We will now sketch this construction and explain how it serves as a basis for a paracomplete theory of truth.

Following Field [3], we use $\{0, 1/2, 1\}$ as the names for the three truth-values, in order to avoid confusion between the object language predicate *True* and the truth-value 1 that was called *true* by Kripke [4]. We assume that $\mathcal{L}^{\text{arithm}}$ contains the falsity constant $\bot$, the negation symbol $\neg$, the conjunction symbol $\wedge$ and the universal quantifier $\forall$. We write $(\varphi \vee \varphi)$ for $\neg(\neg\varphi \wedge \neg\psi)$, $(\varphi \supset \varphi)$ for $\neg(\varphi \wedge \neg\psi)$, and $\exists x : \varphi$ for $\neg\forall x : \neg\varphi$. We sometimes drop brackets when this does not cause confusion. As usual, we assume that $\mathcal{L}^{\text{arithm}}$ contains the equality symbol $=$ as its only predicate symbol and that it contains a constant symbol 0, a unary function symbol *succ* (*successor*) and two binary function symbols $+$ and $\cdot$, conventionally written in infix notation (e.g. $(s(0) \cdot (s(0) + s(0)))$). A countably infinite supply of variable symbols $(x, y, z, x_0, x_1, \dots)$ is assumed to be given. As usual, the constant symbol 0 and the variable symbols can be combined with the function symbols to form *terms*.

A *variable assignment* s is a function that assigns a natural number to each variable. Given a variable assignment s, a variable x and a natural number n,

$s[x:n]$ denotes the variable assignment that coincides with s on all variables other than x and that assigns n to x. One can inductively define the interpretation t^s of a term t under a variable assignment s as follows:

$$\begin{aligned}
0^s &= \text{the natural number } 0 \\
x^s &= \text{the number that } s \text{ assigns to the variable } x \\
succ(t)^s &= \text{the successor of the natural number } t^s \\
(t_1 + t_2)^s &= \text{the sum of the natural number } t_1^s \text{ and the natural number } t_2^s \\
(t_1 \cdot t_2)^s &= \text{the product of the natural number } t_1^s \text{ and the natural number } t_2^s
\end{aligned}$$

Kripke's construction is based on a transfinite recursion which starts with assigning the truth-value ½ to each formula and then recursively updates the truth-values of all formulas until a fixed point is reached after some transfinite number of iterations. At each step α of this transfinite recursion and for each variable assignment s, we assign to each formula $\varphi \in \mathcal{L}_{True}^{\text{arithm}}$ a truth-value $\varphi^{\alpha,s} \in \{0, ½, 1\}$. The transfinite recursion is defined as follows:

$$\begin{aligned}
\varphi^{0,s} &= ½ \text{ for every } \varphi \in \mathcal{L}_{True}^{\text{arithm}} \text{ and every variable assignment } s \\
(t_1 = t_2)^{\alpha+1,s} &= \begin{cases} 1 \text{ if } t_1^s = t_2^s \\ 0 \text{ otherwise} \end{cases} \\
(\bot)^{\alpha+1,s} &= 0 \\
(\neg\varphi)^{\alpha+1,s} &= 1 - \varphi^{\alpha+1,s} \\
(\varphi \wedge \psi)^{\alpha+1,s} &= \min(\varphi^{\alpha+1,s}, \psi^{\alpha+1,s}) \\
(\forall x : \varphi)^{\alpha+1,s} &= \min\{\varphi^{\alpha+1,s[x:n]} \mid n \in \mathbb{N}\} \\
(True(t))^{\alpha+1,s} &= \begin{cases} 1 & \text{if there is a } \varphi \in \mathcal{L}_{True}^{\text{arithm}} \text{ with } t^s = \langle\varphi\rangle \text{ and } \varphi^{\alpha,s} = 1 \\ ½ & \text{if there is a } \varphi \in \mathcal{L}_{True}^{\text{arithm}} \text{ with } t^s = \langle\varphi\rangle \text{ and } \varphi^{\alpha,s} = ½ \\ 0 & \text{otherwise} \end{cases} \\
\varphi^{\lambda,s} &= \begin{cases} ½ & \text{if } \lambda \text{ is a limit ordinal and } \varphi^{\alpha,s} = ½ \text{ for all } \alpha < \lambda \\ 1 & \text{if } \lambda \text{ is a limit ordinal and } \varphi^{\alpha,s} = 1 \text{ for some } \alpha < \lambda \\ 0 & \text{otherwise, if } \lambda \text{ is a limit ordinal.} \end{cases}
\end{aligned}$$

Clearly for sentences (formulas without free variables) the variable assignment has no impact on the assigned truth-value, so we can write φ^α instead of $\varphi^{\alpha,s}$ when φ is a sentence.

One can easily see that this transfinite recursion is monotonic, i.e. if $\varphi^{\alpha,s} \neq ½$ for some ordinal α, then $\varphi^{\beta,s} = \varphi^{\alpha,s}$ for all $\beta \geq \alpha$. This together with the fact that $\mathcal{L}_{True}^{\text{arithm}}$ is countable implies that a fixed point is reached at some countable ordinal α_0, i.e. for each formula $\varphi \in \mathcal{L}_{True}^{\text{arithm}}$, each ordinal $\alpha \geq \alpha_0$ and each variable assignment s, $\varphi^{\alpha,s} = \varphi^{\alpha_0,s}$. The (ultimate) truth-value of a sentence $\varphi \in \mathcal{L}_{True}^{\text{arithm}}$, denoted as $|\varphi|$, is defined to be φ^{α_0}.

The idea behind paracomplete theories of truth like that of Field [3] is that the only sentences of $\mathcal{L}_{True}^{\text{arithm}}$ that we should accept are the ones that get assigned truth-value 1 in Kripke's construction, while sentences with truth-value 0 or ½ should be rejected. The theory comprising all the sentences with truth-value 1 in Kripke's construction is usually called *KFS*.

A sentence φ for which $(\varphi \vee \neg\varphi)$ is accepted is called *bivalent*. All sentences that do not involve the predicate *True* are bivalent in *KFS*. Also any sentence in which *True* is only applied to Gödel codes of sentences not involving *True* is bivalent. This process can be continued to the point that can be informally characterized by saying that all sentences in which there are no infinite nestings of the predicate *True* are bivalent. Note that "n is the Gödel code of a bivalent formula" is itself not generally bivalent, so when we want to bivalently restrict ourselves to bivalent formulas, we need to use a syntactic criterion like "the formula does not contain the predicate *True*" (but this way we always miss out some bivalent formulas).

The three-valued logic that is underlying the theory *KFS* is usually called K_3. Tamminga [8] has defined a natural deduction calculus for K_3, which differs from the standard natural deduction calculus for classical logic only in two respects: While the $\neg$-introduction rule (*proof by contradiction*) is dropped, five rules are added to ensure that we can still perform a sufficient amount of reasoning with negation, one rule for double negation introduction ($\varphi \models \neg\neg\varphi$) and four rules that correspond to the two De Morgan's laws $\neg(\varphi \wedge \psi) \dashv\models (\neg\varphi \vee \neg\psi)$ and $\neg(\varphi \vee \psi) \dashv\models (\neg\varphi \wedge \neg\psi)$. So we can say that the paracomplete approach gives up proof by contradiction while replacing it by weaker reasoning principles.

Due to Gödel's second incompleteness theorem, it is not possible to give a complete recursive axiomatization of the set of sentences of $\mathcal{L}_{True}^{\text{arithm}}$ that get assigned truth-value 1 in Kripke's construction. But a natural axiomatization to use for deriving a considerable subset of these sentences is the one that we get by adding the rules $\varphi \models True(\langle\varphi\rangle)$ and $True(\langle\varphi\rangle) \models \varphi$ as well as the axioms of classical Peano Arithmetic and the axiom scheme "$(\varphi \vee \neg\varphi)$ is an axiom for $\varphi \in \mathcal{L}^{\text{arithm}}$" to the natural deduction calculus for K_3. We call the resulting proof system PA_{KFS}. We write $Pr_{KFS}(n)$ to denote the statement that n is the Gödel code of a formula that can be derived in PA_{KFS}. We say "PA_{KFS} is sound" for the statement $\forall n : (Pr_{KFS}(n) \supset True(n))$.

One can easily see that the Liar sentence L gets assigned truth-value ½ in Kripke's construction, so both L nor its negation $\neg L$ will be rejected in a paracomplete theory of truth.

Usually when we reject a certain statement, we can explain this rejection by explaining that we believe the negation of the statement in question, and maybe additionally give reasons for this belief in the statement's negation. For a defender of a paracomplete theory of truth, this kind of explanation for their rejection of L is not possible, because they also reject $\neg L$. So how could a paracompletist explain their rejection of L?

One thing that they could do is to step outside the object language $\mathcal{L}_{True}^{\text{arithm}}$ and use the metalinguistic vocabulary of Kripke's construction to explain that L does not get truth-value 1 in that construction. But this solution is unsatisfying, because it relies on going to a metalanguage rather than staying within a given language. This immediately raises the question why we don't immediately start with a language (e.g. the language of set theory) in which Kripke's construction can be performed. Actually Field [3] does start with the language of set theory, but in that case the set-class distinction implies that Kripke's construction cannot be performed with $\forall$ interpreted a unrestrictedly quantifying over all sets, but can only be performed with $\forall$ interpreted a quantifying over the members of a fixed set U. And no matter what set U we choose, we always get false sentences that have value 1 with respect to quantification over U. In that case Kripke's construction is not a trustworthy criterion for truth, so reference to it as an explication for one's rejection of a certain sentence is not convincing.

Instead of proposing such a metalinguistic response to the question of how to explain one's rejection of L, Field [3] introduces a determinateness operator D, where $D\varphi$ intuitively means 'determinately φ'. With the help of this operator, Field can say $\neg DL$, i.e. say that the Liar sentence L is not determinately true; this is taken to be a reason for rejecting L that is expressible in the object language. In Field's account, D is not a primitive notion, but is defined in terms of a non-material conditional $\rightarrow$ that Field introduces: $D\varphi$ is taken to mean $\varphi \wedge \neg(\varphi \rightarrow \neg\varphi)$ (or equivalently $\varphi \wedge (\top \rightarrow \varphi)$). The semantics of $\rightarrow$ is explicated through a transfinite revision-rule construction. This is a construction that has some resemblance to Kripke's construction, but it does not have the monotonicity property of that construction. Instead, truth-values of sentences involving $\rightarrow$ can oscillate between different truth-values. If such an oscillation occurs all the way towards a limit ordinal λ, then the truth-value at step λ will be ½.

Once the determinateness operator is introduced, one can form a strengthened Liar sentence L_1 that is provably equivalent to $\neg D\mathit{True}(\langle L_1\rangle)$. This brings up the question what the status of this strengthened Liar sentence is. It turns out that accepting it or its negation would be problematics, but we cannot express rejection of it in the same way as in the case of the Liar sentence, because accepting $\neg DL_1$ would amount to accepting $\neg D\mathit{True}(\langle L_1\rangle)$, i.e. to accepting L_1. What we can do instead is to explain our rejection of L_1 by claiming $\neg DDL_1$, also written as $\neg D^2L_1$. So iterating the determinateness operator yields a stronger notion of determinateness, and this stronger notion can be used to explain our stance towards a sentence involving a weaker notion of determinateness.

But then we can construct a sentence L_2 that is provably equivalent to $\neg D^2\mathit{True}(\langle L_2\rangle)$, and to explain our rejection of L_2 we need an even stronger notion of determinacy, namely D^3. Field shows that this process of iterating D can even be continued into the transfinite, but this involves some technical difficulties that go beyond the scope of this paper. It turns out that the question of how far precisely this can be meaningfully continued into the transfinite is also a very tricky one, as it touches on König's paradox of the least undefinable ordinal [7]. Field observes that for any given hereditarily definable ordinal α (i.e. for any α such that α and all its predecessors are definable), D^α is a definable and well-behaved operator. However, one cannot assume that "α is a hereditarily definable ordinal" is bivalent, because that would lead to a contradiction by König's paradox.

4 A Definable Transfinite Hierarchy of Determinateness Operators

In this section we show how a transfinite hierarchy of determinateness operators can be defined within *KFS*. Before giving the formal definition, let us first start with an informal motivation. We want to be able to say of some formulas that they have a determinate truth-value, namely being determinately true or determinately false, while saying of other formulas that they do not have a determinate truth-value. Additionally, we want this notion of determinateness to have a sensible compositional behavior with respect to the logical connectives and quantifiers, namely:

1. When t_1 and t_2 are variable-free terms that denote the same natural number, then $t_1 = t_2$ is determinately true.
2. When t_1 and t_2 are variable-free terms that denote different natural numbers, then $t_1 = t_2$ is determinately false.
3. $\bot$ is determinately false.
4. When φ is determinately true, $\neg\varphi$ is determinately false.
5. When φ is determinately false, $\neg\varphi$ is determinately true.
6. When φ is determinately false, $\varphi \wedge \psi$ is determinately false.
7. When ψ is determinately false, $\varphi \wedge \psi$ is determinately false.
8. When φ and ψ are both determinately true, $\varphi \wedge \psi$ is determinately true.
9. When $\varphi(t)$ is determinately false, $\forall x : \varphi(x)$ is determinately false.
10. When $\varphi(\bar{n})$ is determinately true for all $n \in \mathbb{N}$, $\forall x : \varphi(x)$ is determinately true. ($\bar{n}$ denotes the term $succ(\ldots succ(0)\ldots)$ with n occurrences of $succ$.)
11. When n is the Gödel code of a determinately true sentence, then $True(n)$ is determinately true.
12. When n is the Gödel code of a determinately false sentence, then $True(n)$ is determinately false.

One can interpret the above criteria for "determinately true" and "determinately false" as an implicit inductive definition of these two notions. If one steps out of the object language of *KFS* and allows a metatheoretic definition, one can transform the inductive definition into an explicit definition: For this, one needs to choose for the extensions of "determinately true" and "determinately false" a pair of sets that satisfies the above criteria such that the sets are minimal with respect to set inclusion among the sets with this property. We cannot quantify over sets of formulas within *KFS*, so we cannot use this strategy to turn the implicit inductive definition into an explicit definition within *KFS*.

Denecker and Vennekens [1] show that various kinds of inductive definitions used in mathematics can be given a unified semantic account with the help of the well-founded semantics of logic programs. In the following we take inspiration from the formal definition of the well-founded semantics to transform the above implicit inductive definition of determinate truth and determinate falsity into an explicit definition of a determinateness operator Δ_1 within *KFS*. Formulas of the form $\Delta_1\varphi$ are not in general bivalent, but whenever they are bivalent, the truth-value of $\Delta_1\varphi$ is identical to the one that gets assigned to the statement "φ is definitely true" in the metatheoretic explicit definition of "determinately true".

In order to explain how our definition is inspired by the well-founded semantics, we give a brief informal sketch of the definition of the well-founded semantics; readers interested in the formal details may consult the paper by Denecker and Vennekens [1]. An *inductive definition* is a set of clauses consisting of a definiendum called *head* and a definiens called *body*. The head is always an atomic formula, and all the predicates that appear in at least one head are considered to be simultaneously defined by this inductive definition. As an example, the above enumerated list can be read as a simultaneous inductive definition of "determinately true" and "determinately false", where each item represents a clause and in each clause, the part between "When" and the comma is the body and the part after the comma is the head (as the clause about $\bot$ shows, the head can also be empty, in which case it is considered to be always true). The well-founded model of an inductive definition can be defined as the limit of a *well-founded induction*, which is a transfinite sequence of approximations to the well-founded model. In each approximation, some atoms involving one of the defined predicates are already known to be true, other such atoms are already known to be false, and others still have unknown truth-value. At each successor step, we refine the previous approximation in one of two possible ways: If our current approximation makes the body of some clause true, we may add the head of that clause to the atoms that have been accepted to be true. And if adding some atoms to the set of atoms considered false results in all bodies that define those atoms to be false, then we may indeed add those atoms to the set of atoms considered false. We continue this process until no more refinement is possible, at which point the well-founded model of the inductive definition has been reached. Note the asymmetry between making atoms true and making atoms false: We are free to assume that atoms are false, as long as this prophecy turns out to fulfill itself, whereas for considering something true at some step, we must have reasons for considering it true already at a previous step.

Inspired by the treatment of falsity in the definition of the well-founded semantics, we want to be able to say that a formula does not have a determinate truth-value if assuming it to not have a determinate truth-value turns out to be a self-fulfilling prophecy. For example, if we assume the Liar sentence L and the sentence $True(L)$ not to have a determinate truth-value, then for $n = \langle L \rangle$ the bodies of the clauses 11 and 12 in the above enumerated list are false, which confirms the non-determinateness of $True(\langle L \rangle)$, which in turn by clause 4 confirms the non-determinateness of $\neg True(\langle L \rangle)$, i.e. of L.

This motivates the definition of the function *conf_ind*, which maps a pair $(\langle\varphi\rangle, \langle\psi(x)\rangle)$ of Gödel codes of formulas to a Gödel code $\langle\chi\rangle = conf\_ind(\langle\varphi\rangle, \langle\psi(x)\rangle)$, where the intuition is that when χ is satisfied then assuming the indeterminateness of all formulas whose Gödel codes satisfy ψ confirms the indeterminateness of φ:

$$\begin{aligned}
conf_ind(\langle t_1 = t_2\rangle, \langle\psi(x)\rangle) &= \langle\bot\rangle \\
conf_ind(\langle\bot\rangle, \langle\psi(x)\rangle) &= \langle\bot\rangle \\
conf_ind(\langle\neg\varphi\rangle, \langle\psi(x)\rangle) &= \langle\psi(\langle\varphi\rangle)\rangle \\
conf_ind(\langle\varphi_1 \wedge \varphi_2\rangle, \langle\psi(x)\rangle) &= \langle(\psi(\langle\varphi_1\rangle) \vee Pr_{KFS}(\varphi_1)) \wedge (\psi(\langle\varphi_2\rangle) \vee Pr_{KFS}(\varphi_2)) \\
&\quad \wedge (\psi(\langle\varphi_1\rangle) \vee \psi(\langle\varphi_2\rangle))\rangle \\
conf_ind(\langle\forall x : \varphi(x)\rangle, \langle\psi(x)\rangle) &= \langle\forall n : (\psi(\langle\varphi(\bar{n})\rangle) \vee Pr_{KFS}(\langle\varphi(\bar{n})\rangle)) \wedge \exists n : \psi(\langle\varphi(\bar{n})\rangle)\rangle \\
conf_ind(\langle True(n)\rangle, \langle\psi(x)\rangle) &= \langle\psi(n)\rangle
\end{aligned}$$

The following lemma, which one can easily derive from the definition of *conf_ind* and Kripke's construction, formalizes the idea behind the intuitive meaning of $conf\_ind(\langle\varphi\rangle, \langle\psi(x)\rangle)$ mentioned above:

Lemma 1. *Let $\varphi, \psi \in \mathcal{L}^{\text{arithm}}_{True}$ and let α be an ordinal. If PA_{KFS} is sound, $|conf\_ind(\langle\varphi\rangle, \langle\psi(x)\rangle)| = 1$ and for all formulas χ such that $|\psi(\langle\chi\rangle)| = 1$, we have $\chi^\alpha = 1/2$, then $\varphi^{\alpha+1} = 1/2$.*

Now we use this function *conf_ind* to define the predicate Ind_1, where the intuitive meaning of $Ind_1(\langle\chi\rangle)$ is that χ has indeterminate truth-value. $Ind_1(\langle\chi\rangle)$ is defined to be the *KFS* formalization of the statement "There exists a number n that is the Gödel code of a formula $\psi(x)$ such that $\psi(x)$ does not contain the predicate *True*, $\psi(\langle\chi\rangle)$ is true and for any number m, if $\psi(m)$ is true then the formula whose Gödel code is $conf\_ind(m, n)$ is true." Intuitively, this definition says that there are some formulas (namely those whose Gödel codes satisfy $\psi(x)$) that contain χ and that have the property that assuming them to be indeterminate confirms their indeterminateness. For choosing a collection of formulas that we assume to be indeterminate, we make use of a predicate $\psi(x)$ that does not contain *True*. The fact that it does not contain *True* ensures that it is bivalent. If we allowed for an arbitrary (possibly non-bivalent) formula at this place, we would never be able to accept $\neg Ind_1(\chi)$ for any χ, because a non-bivalent $\psi(x)$ will ensure that the truth-value of $Ind_1(\chi)$ in Kripke's construction is at most $1/2$.

The following lemma formalizes the idea that $Ind_1(\langle\varphi\rangle)$ expresses the indeterminateness of φ:

Lemma 2. *If PA_{KFS} is sound and $|Ind_1(\langle\varphi\rangle)| = 1$, then $|\varphi| = 1/2$.*

Proof. The soundness of PA_{KFS}, the definition of Ind_1 and the fact that $|Ind_1(\langle\varphi\rangle)| = 1$ together imply that there exists a formula $\psi(x)$ such that $\psi(x)$ does not contain the predicate *True*, $|\psi(\langle\varphi\rangle)| = 1$ and for any $\chi \in \mathcal{L}^{\text{arithm}}_{True}$, if $|\psi(\langle\chi\rangle)| = 1$ then $|conf\_ind(\langle\chi\rangle, \langle\psi(x)\rangle)| = 1$. Let Γ be the set of all $\chi \in \mathcal{L}^{\text{arithm}}_{True}$ such that $|\psi(\langle\chi\rangle)| = 1$.

By a transfinite induction one can prove that for every α and every $\chi \in \Gamma$, $\chi^\alpha = 1/2$. The inductive step directly follows from Lemma 1. Since $|\psi(\langle\varphi\rangle)| = 1$, we have that $\varphi \in \Gamma$, i.e. that $\varphi^\alpha = 1/2$ for all α, as required. □

Now we define the determinateness operator Δ_1 as follows: $\Delta_1\varphi$ is defined to be shorthand notation for $\varphi \wedge \neg Ind_1(\langle\varphi\rangle)$. The following theorem, which one can easily derive from the definition of Δ_1 and Lemma 2, ensures that $\neg\Delta_1\varphi$ can be used to explain one's rejection of φ.

Theorem 1. *If PA_{KFS} is sound and $|\neg\Delta_1\varphi| = 1$, then $|\varphi| = 0$ or $|\varphi| = 1/2$.*

Now using this determinateness operator Δ_1, we can explain our rejection of L by saying $\neg\Delta_1 L$. In order to see that $|\neg\Delta_1 L| = 1$, note that the definition of *conf_ind* implies that $|conf\_ind(\langle True(L)\rangle, \langle x = \langle L\rangle \vee x = \langle True(L)\rangle\rangle)| = 1$. The standard construction of L together with the definition of *conf_ind* furthermore implies that $|conf\_ind(\langle L\rangle, \langle x = \langle L\rangle \vee x = \langle True(L)\rangle\rangle)| = 1$. These two facts together with the definition of Ind_1 imply that $|Ind_1(\langle L\rangle)| = 1$, i.e. that $|\neg\Delta_1 L| = 1$.

Once we have successfully dealt with the Liar sentence L in this way, the obvious next question to ask is what happens to a strengthened version of the Liar that makes use of Δ_1. One can construct a strengthened Liar sentence L_1' that is provably equivalent to $\neg\Delta_1 True(\langle L_1'\rangle)$. Similarly to the case of the strengthened Liar sentence L_1 based on Field's determinateness operator D, one can show that $|L_1'| = 1/2$, but that the rejection of L_1' cannot be explained in the same way as the rejection of L, because $|\neg\Delta_1 L_1| = |\neg\Delta_1 True(\langle L_1'\rangle)| = |L_1'| = 1/2$. Unlike in the case of Field's approach, one cannot get around this problem by iterating the determinateness operator, because if $|\neg\Delta_1\Delta_1 L_1'|$ were 1, then there would be a witness $\psi(x)$ for $Ind_1(\langle\Delta_1 L_1'\rangle)$, in which case $\psi'(x) := (x = \langle L_1'\rangle \vee x = \langle\Delta_1 True(\langle L_1'\rangle)\rangle \vee \psi(x))$ would be a witness for $Ind_1(L_1')$, which would be a contradiction.

What one can do instead is to define a stronger determinateness operator Δ_2: $\Delta_2\varphi$ is shorthand for $\varphi \wedge \neg Ind_2(\langle\varphi\rangle)$, where Ind_2 is defined just like Ind_1 with the only difference being that the formula $\psi(x)$ is not required to lack the predicate *True*, but instead has to satisfy the criterion that *True* is only applied to Gödel codes of sentences not involving *True*. By choosing $\psi(x)$ to be $Ind_1(x)$, we can establish that $|Ind_2(L_1')| = 1$, i.e. that $|\neg\Delta_2 L_1'| = 1$. So we can use the stronger determinateness operator Δ_2 to explain our rejection of L_1'.

Similarly one can construct a determinateness operator Δ_3 based on a predicate Ind_3 in which the formula $\psi(x)$ may apply the predicate *True* only to formulas that apply *True* only to *True*-free formulas. In order to define these determinateness operators further, we need the following depth predicate:

- We say $depth(1, n)$ if n if the Gödel code of a formula in $\mathcal{L}^{\text{arithm}}$.
- We say $depth(\alpha + 1, n)$ if α is an ordinal notation and n is the Gödel code of a formula in which *True* is only applied to a term t if t has been restricted by a syntactic criterion that implies $depth(\alpha, t)$.
- We say $depth(\lambda, n)$ if λ is a limit ordinal notation and n is the Gödel code of a formula that satisfies $depth(\alpha, n)$ for all $\alpha < \lambda$.

Now we can define Ind_α for any ordinal notation α by modifying the above definition of Ind_1 by allowing ψ to be any formula that satisfies $depth(\alpha, \langle\psi\rangle)$, and we can define $\Delta_\alpha\varphi$ to be shorthand for $\varphi \wedge \neg Ind_\alpha(\langle\varphi\rangle)$.

This defines a transfinite hierarchy of ever stronger determinateness operators. Similarly as in the case of the transfinite iterations of Field's determinateness operator, we can explain the rejection of a strengthened Liar sentence that uses a determinateness operator by using a stronger determinateness operator.

5 Conclusion

Just like Field [3], we have defined a determinateness operator to explain the rejection of the Liar sentence within the object language of our theory of truth. Unlike Field's determinateness operator, our determinateness operator can be defined within the theory *KFS* and does not require *KFS* to be extended by a conditional that is undefinable within *KFS*. Field's approach, on the other hand, is based on the idea of extending *KFS* through a semantic construction that involves the combination of a revision-rule construction for the semantics of the conditional $\rightarrow$ and Kripke's construction for the semantics of *True*, so the overall semantic construction is rather complicated. Additionally, he does not provide a proof-theoretic characterization of the ensuing theory. Our approach, on the other hand, can be completely developed within the theory *KFS* that comes out of Kripke's construction and that has a natural proof theory. Thus we avoid the complications of the extended theory for the conditional that Field has developed while achieving an equally good resolution of the Liar paradox and strengthened versions of it.

Unlike Field's determinateness operator, our determinateness operator cannot be strengthened by iterating it. Instead, we have defined a transfinite hierarchy of ever stronger determinateness operators that are not definable as iterations of the weakest determinateness operator. With this transfinite hierarchy of determinateness operators we can explain the Liar paradox and strengthened versions of it in much the same way as Field does, only that we use our stronger determinateness operators where Field uses iterations of his determinateness operator.

Field motivates his extension of *KFS* to a theory with a conditional not only through the fact that this allows him to express his rejection of the Liar sentence in the object language, but also based on the argument that it is desirable to have a conditional that satisfies certain basic properties that one would expect of a conditional but that are not satisfied by the material implication $\supset$ of *KFS*, e.g. that $\varphi \rightarrow \varphi$ and $\varphi \rightarrow (\varphi \vee \psi)$ are logical truths for any choice of φ and ψ. A thorough discussion of this issue would go beyond the scope of this paper, but let me very briefly sketch how the formal apparatus developed in this paper could also be used to define a conditional $\varphi \Rightarrow \psi$ that has such desirable properties.

For this purpose we define a predicate $Cond\_Ind(m, n)$ for conditional indeterminateness as follows: $Cond\_Ind(\langle\varphi\rangle, \langle\chi\rangle)$ is defined to be the *KFS* formalization of the statement "There exists a number n that is the Gödel code of a formula $\psi(x)$ such that $\psi(x)$ does not contain the predicate *True*, $\psi(\langle\varphi\rangle)$ is true, $\psi(\langle\chi\rangle)$ is true and for any number $m \neq \langle\varphi\rangle$, if $\psi(m)$ is true then the formula whose Gödel code is $conf\_ind(m, n)$ is true." Using this predicate, we define

$\varphi \Rightarrow \psi$ to be shorthand for $(\varphi \supset \psi) \vee Cond\_Ind(\langle\varphi\rangle, \langle\psi\rangle)$. One can easily verify that this conditional satisfies modus ponens and that $\varphi \Rightarrow \varphi$ and $\varphi \Rightarrow (\varphi \vee \psi)$ are true for any choice of φ and ψ. The further exploration of this conditional is left to future work.

References

1. Denecker, M., Vennekens, J.: The Well-Founded Semantics Is the Principle of Inductive Definition, Revisited. In: Principles of Knowledge Representation and Reasoning: Proceedings of the Fourteenth International Conference, KR 2014, Vienna, Austria, July 20-24, 2014 (2014)
2. Feferman, S.: Reflecting on incompleteness. The Journal of Symbolic Logic **56**(1), 1–49 (1991)
3. Field, H., et al.: Saving truth from paradox. Oxford University Press (2008)
4. Kripke, S.: Outline of a theory of truth. The journal of philosophy **72**(19), 690–716 (1976)
5. Priest, G., et al.: In contradiction. Oxford University Press (2006)
6. Ripley, D.: Comparing substructural theories of truth. Ergo, an Open Access Journal of Philosophy **2** (2015)
7. Simmons, K.: Paradoxes of denotation. Philosophical Studies: An International Journal for Philosophy in the Analytic Tradition **76**(1), 71–106 (1994)
8. Tamminga, A.: Correspondence analysis for strong three-valued logic. Логические исследования (20) (2014)
9. Van Gelder, A., Ross, K.A., Schlipf, J.S.: The well-founded semantics for general logic programs. Journal of the ACM (JACM) **38**(3), 619–649 (1991)

Paradoxes behind the Solovay sentences *

Ming Hsiung[1][0000−0001−9037−2024]

South China Normal University, Guangzhou, Guangdong 510631, China
mingshone@163.com

Abstract. The sentences that Solovay constructed in his famous theorem on arithmetical completeness of Gödel-Löb provability logic are all undecidable. We use Solovay's method to construct paradoxes, which bear to the Solovay's sentences much the same relation as the liar paradox bears to the Gödel sentence. The main idea is to use the truth predicate instead of the provability predicate in the formalisation of the Solovay function. A typical example of such paradoxes may be seen as obtained from two ordinary paradoxes by damaging symmetry of the 'baptising' biconditionals. We prove that this paradox is a proper weakening of the latter two in the sense that the former has a strictly lower degree of paradoxicality than the latter two. Solovay's method provides a new approach to finding various kinds of paradoxes.

Keywords: Paradox· Provability· Solovay sentences· Truth

1 Introduction

To prove his famous theorem on arithmetical completeness of Gödel-Löb provability logic GL, Solovay [6] introduced the sentences, each of which asserts a specific function of arithmetic converges to a number. Like the sentence that Gödel constructed to prove his first incompleteness theorem, the Solovay sentences are all undecidable, and their undecidability are closely related to the features of a 'standard' provability predicate. Since Gödel sentence rises from the liar paradox, we wonder whether there are paradoxes lying behind the Solovay sentences.

The Solovay sentences are given with respect to a finite relational frame $\langle W, R\rangle$ (denoted by $\mathcal{F}$ henceforth).[1] W is a set of natural numbers containing

* Supported by National Social Science Foundation of China (grant number 19BZX136). The author would like to thank three anonymous referees for their comments. These comments were very useful for improving this paper. This chapter is in its final form and it is not submitted to publication anywhere else.

[1] Solovay's theorem on GL says that a formula is provable in GL, iff any arithmetical translation of it is also provable in Peano arithmetic (only the right-to-left direction is non-trivial). For the proof, a preliminary result is that a formula is provable in GL, iff it is valid in all finite transitive and conversely well-founded frames. Thus, if a formula is not provable in GL, it must be falsified by some finite transitive and conversely well-founded Kripke model. We obtain the desired arithmetical transla-

B. Liao and Y. N. Wáng (eds.), *Context, Conflict and Reasoning*,
Logic in Asia: Studia Logica Library,
https://doi.org/10.1007/978-981-15-7134-3_2

at least 0. Each world i of W corresponds to a Sololvay sentence S_i. S_i, as just mentioned, is a sentence asserting that a certain number-theoretic function has the limit i. The definition of the function, namely h, in turn depends upon the Sololvay sentences: $h(0) = 0$ and for any $m \geq 0$, if $h(m)Ri$ and $\mathrm{Pf}(m, \ulcorner \neg S_i \urcorner)$, then $h(m+1) = i$, otherwise $h(m+1) = h(m)$. Here, Pf is a standard proof predicate of PA, which holds for the numbers m_1 and m_2, iff m_1 is the Gödel number of a proof in Peano arithmetic (PA) of a sentence whose Gödel number is m_2. Seeing the construction, we may naturally try to replace the proof predicate by the truth predicate, $T(x)$, and see whether the sentences we get are paradoxical.[2] This is the main idea of bringing the paradoxes up out of the Solovay sentences.

To be more precise, we are considering the sentences σ_i ($i \in W$), asserting the following function h has the limit i: $h(0) = 0$ and for all $m \geq 0$,

$$h(m+1) = \begin{cases} i, & \text{if } h(m)Ri \text{ and } \neg T \ulcorner \sigma_i \urcorner; \\ h(m), & \text{otherwise.} \end{cases} \tag{1}$$

It is well-known that the binary relation R of $\mathcal{F}$ is supposed to be transitive and conversely well-founded in the construction of the Solovay sentences. For now, we only need to suppose that $\mathcal{F}$ is a finite frame. As we will see, it suffices to formalise h and so construct the new sentences σ_i in PA. Note also that the range of h is the set $\{i \mid 0R^*i\}$, where R^* is the reflexive transitive closure of R, that is, iR^*j, iff $i = i_0 Ri_1 R \dots Ri_k = j$ for some $i_0, \dots, i_k$ in W (and in particular, $i = j$ when $k = 0$). Thus, without loss of generality, we can suppose further $\mathcal{F}$ satisfies: $0R^*i$ for all $i \in W$. Otherwise, we can consider the sub-frame of $\mathcal{F}$ generated by the point 0 instead of $\mathcal{F}$ itself.

As we will see, whether the sentences σ_i are paradoxical is dependent on something like the converse well-foundedness of the frame. The main result of this essay is that for any frame $\mathcal{F}$ containing more than one point, the sentences σ_i are paradoxical, iff every directed cycle in $\mathcal{F}$ (if any) has only the length 1. In particular, whenever $\mathcal{F}$ is conversely well-foundedness (and at-least-two-points), the sentences σ_i are paradoxical. It provides a new approach to constructing paradoxes.

A highlight of our construction is that by the present approach, we can construct the paradoxes that are radically different from all the known paradoxes. All the known paradoxical sentences have the perfect symmetry in the sense that each of them can be directly characterised by a biconditional. By contrast, among the paradoxical sentences we construct, some of them cannot be characterised by biconditionals: the best characterising conditions for them are merely the

tion by embedding the Kripke model into Peano arithmetic. It is at this point that Solovay's ingenious function plays a primary role in the proof of Solovay's theorem. We refer the reader to [1, pp. 125-131] or [5, pp. 134-142] for a full exposition of the proof.

[2] A more symmetrical replacement is to use the satisfaction predicate, which is also binary. For simplicity, we here insist on employing the truth predicate rather than the satisfaction predicate.

conditionals. In some sense, such sentences can be seen as obtained from the ordinary ones by damaging symmetry of characterising biconditionals. This is the main point that we will articulate in this paper.

2 Construction

Let $\mathscr{L}_T$ be the language obtained from the first-order arithmetic language by adding a unitary predicate symbol T. Let PA denote the first-order arithmetic, and PAT denote the theory which is the same as PA except that the induction schema can be applied to the formulas containing T. By a formula (sentence), we always mean a formula (sentence) in $\mathscr{L}_T$. Unless otherwise claimed, to say a formula is provable is to say it is provable in PAT.

For any set X of natural numbers and for a sentence A, we use $\langle \mathfrak{N}, X \rangle \models A$ to denote that A is true in the expanded model of the standard structure $\mathfrak{N}$ of natural numbers, in which X is the extension of T. Since $\mathfrak{N}$ is the ground model for PA throughout this paper, we will use $X \models A$ instead of $\langle \mathfrak{N}, X \rangle \models A$. Note that the mathematical induction holds in our meta-language, that is, for any set X of natural numbers, if 0 belongs to X and for any natural number n, n belonging to X implying its successor belonging to X, then all natural number belongs to X. It follows that $X \models \mathrm{PAT}$ for any set X of natural numbers.

We can code the expressions of $\mathscr{L}_T$ by a standard Gödel numbering. We use $\ulcorner A \urcorner$ for the Gödel number of A. Note that for any number n, the corresponding numeral is denoted by $\mathbf{n}$. Nevertheless, for convenience, we identify a number with the corresponding numeral when this number codes a formula. Thus, when A is a sentence, $T\ulcorner A \urcorner$ (i.e., $T(\ulcorner A \urcorner)$) is a well-formed formula whose intended meaning is that A is true.

About the truth predicate, the most well-known result is Tarski's theorem of undefinability of truth. One version of this theorem is that no extension X of T can satisfy all instances of Tarski's T-schema $T\ulcorner A \urcorner \leftrightarrow A$. To see this, we can construct a sentence, namely λ, such that $\mathrm{PA} \vdash \lambda \leftrightarrow \neg T\ulcorner \lambda \urcorner$. The construction is the so-called Gödel diagonalisation. We will use this method to formulate the function h in equation (1). We refer the reader to [5, p. 44] or [1, p. 53] for more details. Now, we can see that for any extension X of T, $X \models \lambda \leftrightarrow \neg T\ulcorner \lambda \urcorner$, and it follows that X cannot satisfy the instance $T\ulcorner \lambda \urcorner \leftrightarrow \lambda$. In this sense, we say the sentence λ is paradoxical—it is well-known as 'the liar paradox.' Generally, we have the following folklore definition of paradoxicality.

Definition 1. *A set Σ of sentences is paradoxical, if no extension X of T can be such that $X \models T\ulcorner A \urcorner \leftrightarrow A$ for all $A \in \Sigma$.*

Throughout the paper, we suppose that $\mathcal{F} = \langle W, R \rangle$ is a frame, whose domain is a finite subset of natural numbers. Without loss of generality, let $W = \{0, 1, \dots, n\}$. Moreover, we also suppose that the binary relation is such that $0R^*i$ for all $i \in W$.

The sentences we are going to construct are a slight variant of Solovay's original ones. We will follow the construction that Boolos presented in his book [1, pp. 126ff]. The following notations are useful. FinSeq(x) is the standard formula that defines the predicate 'x is (the Gödel code of) a finite sequence of natural numbers.' lh(x) is the term whose value for $x = m$ is the length of the sequence coded by m. At last, $(x)_y$ is the term whose value for $x = m$ and $y = i$ is the i-th element of the sequence coded by m (ibid., p. 37).

We start the construction by formalising the function h in equation (1). For variables x_1 and x_2, $\lim(x_1, x_2)$ be a term denoting a function whose value for $x_1 = \ulcorner A(v_0, v_1) \urcorner$ and $x_2 = i$ is the Gödel number of the formula

$$\exists v_2 \forall v_0 \geq v_2 \exists v_1 (v_1 = \mathbf{i} \wedge A(v_0, v_1)).$$

Let $B(v_0, v_1, v_2)$ be the formula:

$$\begin{aligned} &\exists v_3 \,(\mathrm{FinSeq}(v_3) \wedge \mathrm{lh}(v_3) = v_0 \wedge (v_3)_0 = \mathbf{0} \wedge (v_3)_{v_0} = v_1 \\ &\wedge \forall v_4 < v_0 \wedge_{i:i \in W} \,((v_3)_{v_4} = \mathbf{i} \rightarrow \wedge_{j:iRj} \,(\neg T \ulcorner \lim(v_2, \mathbf{j}) \urcorner \rightarrow (v_3)_{v_4+\mathbf{1}} = \mathbf{j}) \\ &\wedge (\wedge_{j:iRj} T \ulcorner \lim(v_2, \mathbf{j}) \urcorner \rightarrow (v_3)_{v_4+\mathbf{1}} = \mathbf{i}))) \end{aligned} \qquad (2)$$

Note that there are three conjunctions in equation (2). They are well-formed because the domain of $\mathcal{F}$ is finite.

By diagonalisation, we can find a formula $H(v_0, v_1)$ such that

$$H(v_0, v_1) \leftrightarrow B(v_0, v_1, \ulcorner H(v_0, v_1) \urcorner)$$

is provable (in PAT). At last, for all $i \in W$, we define σ_i to be the sentence

$$\exists v_2 \forall v_0 \geq v_2 \exists v_1 (v_1 = \mathbf{i} \wedge H(v_0, v_1)).$$

Let us emphasise that in this construction, we give a sentence, σ_i, for any point i in the domain of $\mathcal{F}$. $H(v_0, v_1)$ is essentially a fixed point of the predicate shown in equation (2), which is the formalisation of the function h in equation (1). The thought behind h is that h, in Smoryński's terms, is 'reluctantly attempting to climb its way through R by starting at 0 and only moving to an accessible node' when it is true that h will not stay there.[3]

The following are the basic properties of sentences σ_i.

(1) For all $i, j \in W$ with $i \neq j$, $\sigma_i \rightarrow \neg \sigma_j$ is provable.
(2) For all $i, j \in W$ with $i \neq j$ and iRj, $\sigma_i \rightarrow T \ulcorner \sigma_j \urcorner$ is provable.
(3) For all $i \in W$ with $i \neq 0$, $\sigma_i \rightarrow \neg T \ulcorner \sigma_i \urcorner$ is provable.

The first result follows from the fact that $\forall v_0 \exists ! v_1 H(v_0, v_1)$ is provable. The second can be obtained by formalising the following argument in PAT: σ_i implies that there exists m such that $\forall v_0 \geq \mathbf{m} H(v_0, \mathbf{i})$. Assume $\neg T \ulcorner \sigma_j \urcorner$, then by iRj

[3] Cf. [5, p. 136]. Of course, Smoryński's construction is made for Solovay's original function, which involves the provability predicate rather than the truth predicate. Here, we use 'when it is true that' instead of Smoryński's words 'when it is proven that.'

and definition of H, we know $H(\mathbf{m}+\mathbf{1},\mathbf{j})$. Also, we know $H(\mathbf{m}+\mathbf{1},\mathbf{i})$. This is impossible because $i \neq j$. At last, for the third result, only note that whenever $i \neq 0$, $H(v_0, \mathbf{i}) \to \neg T\ulcorner\sigma_i\urcorner$ is provable.

Note that $\vee_{i\in W}\sigma_i$ is not necessarily provable. This is because the function h may not have a limit at all when R is conversely ill-founded. This point will become apparent below.

It is clear that the sentences σ_i are determined by the frame $\mathcal{F}$. We will use a coinage 'Solovayesque' to describe these sentences so that it is not confused with the original Solovay sentences.

3 Main Theorem

In this section, we study whether the Solovayesque sentences on a (finite) frame are paradoxical. Our target is to prove Theorem 1.

Lemma 1. *Suppose $\mathcal{F}$ is a frame whose domain contains at least a world besides 0. Then the set of Solovayesque sentences on $\mathcal{F}$, $\{\sigma_i \mid i \in W\}$, is paradoxical, provided that 'H has a limit' is true, i.e., for all set X of natural numbers, $X \models \vee_{i\in W}\,\sigma_i$.*

Proof. Assume there exists a set X of natural numbers, such that for any $i \in W$, $X \models T\ulcorner\sigma_i\urcorner \leftrightarrow \sigma_i$. By basic property (3), we can see that for all $i \neq 0$, $X \models \sigma_i \to \neg\sigma_i$, that is, $X \models \neg\sigma_i$. By our supposition of the lemma, $X \models \vee_{i\in W}\,\sigma_i$. Thus, $X \models \sigma_0$. What is more, there is a number $i_0 \in W$ with $i_0 \neq 0$. Since $0R^*i_0$, by property (2), we can obtain $X \models \sigma_0 \to \sigma_{i_0}$. Hence, $X \models \sigma_{i_0}$. But we just proved $X \models \neg\sigma_{i_0}$, a contradiction. In conclusion, Σ must be paradoxical.

Lemma 2. *If $\mathcal{F}$ is a conversely well-founded frame whose domain contains at least two points, then 'H has a limit' is true.*

Proof. First, $H(v,\mathbf{i}) \to \vee_{j:iR^*j}\sigma_j$ is provable. Noticing that the converse of R is well-founded, we can prove this result by induction on the converse of R. We omit the details. Now since $H(\mathbf{0},\mathbf{0})$ is provable, and $0R^*i$ holds for all $i \in W$, we know $\vee_{i\in W}\sigma_i$ is provable (Cf. the result (4) in [1, p. 129]). It follows that 'H has a limit' is true.

An immediate consequence of Lemma 1 and 2 is as follows:

Proposition 1. *If $\mathcal{F}$ is a conversely well-founded frame whose domain contains at least two points, then the set of Solovayesque sentences on $\mathcal{F}$ is paradoxical.*

In the above result, we require that the domain of $\mathcal{F}$ contains more than one point. This condition is necessary, because if the domain of $\mathcal{F}$ is a singleton frame, then no matter whether R is conversely well-founded or not, the set $\{\sigma_0\}$ is not paradoxical. In this case, h is actually the zero function, or formally, $\forall v H(v,\mathbf{0})$ is provable. And so is σ_0. Let X be the set of (the Gödel number of) σ_0, then it is evident that $X \models T\ulcorner\sigma_i\urcorner \leftrightarrow \sigma_i$ for $i \in W$.

In the frame $\langle W, R\rangle$, let us say a finite sequence of points in W, say, $i_0, i_1, ..., i_k$, is a directed cycle, if these points are pairwise different except $i_0 = i_k$, and for all $0 \leq j < k$, $i_j R i_{j+1}$. k is the length of this directed cycle. Clearly, if a finite frame is conversely ill-founded, then there must be some directed cycle in it.

Proposition 2. *If $\mathcal{F}$ is a conversely ill-founded frame whose domain contains at least two points, then the set of Solovayesque sentences on $\mathcal{F}$ is paradoxical, iff every directed cycle in $\mathcal{F}$ has the length 1.*

Proof. For the necessity, suppose there is at least a directed cycle of length greater than 1 in $\mathcal{F}$. let us fix one, say, a directed cycle $i_0 R i_1 R ... R i_k$ $(i_k = i_0)$ for some $k > 0$. We consider two cases.

First, suppose 0 occurs in this cycle. In this case, without loss of generality, let $i_0 = 0$. Consider the function h such that for any $a \in \mathbb{N}$, if $a = j(\text{mod } k)$ for some j with $0 \leq j < k$, then $h(a) = i_j$. Clearly, h has no limit. Formalise h by the predicate H in the language for PAT. For $i \in W$, let σ_i be the corresponding Solovayesque sentences. Now fix $X = \varnothing$, we can easily see $X \models T\ulcorner\sigma_i\urcorner \leftrightarrow \sigma_i$ for all i of W.

Second, suppose 0 does not occur in the above cycle. Then there exist numbers $l_0, l_1, \ldots l_m$ $(m \geq 0)$ such that $l_0 = 0$, $l_0 R \ldots R l_m$, $l_m R i_j$ for some $0 \leq j < k$, and there are no common points between $l_0, l_1, \ldots l_m$ and $i_0, i_1, \ldots, i_{k-1}$. Without loss of generality, suppose $l_m R i_0$. Define a function h such that $h(a) = l_a$ for all $0 \leq a \leq m$, and $h(a) = i_j$ for all $a > m$ with $a - (m+1) = j(\text{mod } k)$. Then as above, we can see the set $\{\sigma_i \mid i \in W\}$ is not paradoxical.

The proof of the sufficiency is a reduction to Proposition 1. let $\mathcal{F}$ have and only have a directed cycle of length 1. The main idea is to find a conversely well-founded and more-than-one-point frame $\mathcal{F}'$ together with the corresponding predicate H', such that 'H' has a limit' is true, iff 'H has a limit' is true. Then we can obtain the sufficiency by Lemma 1 and Lemma 2.

Call a point i of W is a *loop end*, if for all $j \in W$, iRj, iff $i = j$. Let R^- be the relation obtained from R by removing all pairs $\langle i, i\rangle$ such that i is a loop end of W, and let $\mathcal{F}^-$ be the frame $\langle W, R^-\rangle$. Clearly, 'H, being a diagonal predicate on $\mathcal{F}^-$, has a limit' is true, iff 'it, being a diagonal predicate on $\mathcal{F}$, has a limit' is true.

Next, call a point i of W is a *loop median*, if iRi and for some $j \neq i$, iRj. If there is a loop median, say i, in the frame $\mathcal{F}^-$, the function h in equation (1) may have the value i at finite consecutive arguments. That is, for some positive integer m and some natural number l_0, $h(x) = i$ for all x with $l_0 \leq x < l_0 + m$. In that case, we can obtain a frame $\mathcal{F}^+$ from $\mathcal{F}^-$ by replacing world i with m many new points, say $i_0, \ldots, i_{m-1}$. Specifically, the domain of $\mathcal{F}^+$ is obtained from W by removing i and adding i_k $(0 \leq k < m)$. Moreover, the binary relation of $\mathcal{F}^+$ is obtained from R^- by removing all pairs $\langle i, j\rangle$ with iR^-j and $\langle j, i\rangle$ with jR^-i, and adding $\langle i_k, i_{k+1}\rangle$ for all k such that $0 \leq k < m$, $\langle j, i_0\rangle$ for all j such that jR^-i, and $\langle i_k, j\rangle$ for all j such that iR^-j.

Now define a function h^+ such that h^+ is the same as h except that $h^+(l_k) = i_k$ for $0 \leq k < m$. Clearly, 'H^+ has a limit' is true, iff 'H does so' is also true.

Repeating the above process, we can remove all loop medians of $\mathcal{F}$ one by one, and eventually obtain a conversely well-founded frame, which is denoted by $\mathcal{F}'$. The function h' defined on $\mathcal{F}'$ has the same convergence as the original function h. Thus, we get the desired frame $\mathcal{F}'$, and the corresponding predicate H' such that 'H' has a limit' is true, iff 'H does so' is also true.

To sum up, we get the following result:

Theorem 1. *let $\mathcal{F}$ be a frame, then*

(1) *if $\mathcal{F}$ is a frame containing only the point 0, then the set of all Solovayesque sentences on $\mathcal{F}$ is not paradoxical;*
(2) *otherwise, the set of all Solovayesque sentences on $\mathcal{F}$ is paradoxical, iff every directed cycle in $\mathcal{F}$ (if any) has only the length 1.*

4 Examples

Theorem 1 gives a way of constructing paradoxical sentences. To see how our construction works, we give a few examples in this section.

Example 1. The Solovayesque sentences on $\mathcal{F}$, in which the domain is $\{0, 1\}$, and the binary relation is $\{\langle 0, 1\rangle\}$.

Let σ_0 and σ_1 be the two Solovayesque sentences on $\mathcal{F}$. Then $\sigma_0 \vee \sigma_1$, $\sigma_0 \to \neg\sigma_1$, $\sigma_0 \to T\ulcorner\sigma_1\urcorner$, and $\sigma_1 \to \neg T\ulcorner\sigma_1\urcorner$ are all provable. By Proposition 1, the set $\{\sigma_0, \sigma_1\}$ is paradoxical. As we will see immediately, σ_0 and σ_1 are indeed equivalent to two well-known paradoxical sentences. Note that the first two of the provable sentences we just mentioned implies $\sigma_0 \leftrightarrow \neg\sigma_1$ is also provable. The third one together the biconditional we just get implies $\neg T\ulcorner\sigma_1\urcorner \to \sigma_1$ is provable, and so is $\neg T\ulcorner\sigma_1\urcorner \leftrightarrow \sigma_1$. Hence, σ_1 is precisely equivalent to the liar sentence, and σ_1 is equivalent to the negation of σ_0.

Remind of the reader that both σ_0 and σ_1 are the sentences saying some arithmetic function converges to a number. Neither of them directly says its untruth. But it is proved that they are equivalent to the liar sentence and the negation of the liar respectively. σ_0 and σ_1 may be seen as some 'real' arithmetic sentences, even though they are not so 'pure' as the statement 'one plus two is equal to three' (as their definitions are indeed closely related to the truth predicate). Thus, some 'real' arithmetic sentences are equivalent to the self-referential sentences that are usually seen as purely 'logical.'

Example 2. The Solovayesque sentences on $\mathcal{F}$, in which the domain is $\{0, 1\}$ and the binary relation is $\{\langle 0, 1\rangle, \langle 1, 0\rangle\}$.

In this case, we have $\sigma_0 \to \neg\sigma_1$, $\sigma_0 \to T\ulcorner\sigma_1\urcorner$, $\sigma_1 \to T\ulcorner\sigma_0\urcorner$, and $\sigma_1 \to \neg T\ulcorner\sigma_1\urcorner$ are provable. Nevertheless, $\sigma_0 \vee \sigma_1$ is not provable. For this, let the extension of T be the empty set, then for each even number n, $\varnothing \models H(\mathbf{n}, 0)$, and for each odd number n, $\varnothing \models H(\mathbf{n}, \mathbf{1})$. That means neither $\varnothing \models \sigma_0$ nor $\varnothing \models \sigma_1$.

Hence, $\varnothing \models T\ulcorner\sigma_i\urcorner \leftrightarrow \sigma_i$, $i = 0, 1$. The empty set thus witnesses that $\{\sigma_0, \sigma_1\}$ is not paradoxical. It is also worth pointing out that σ_0 and σ_1 are something like the double truth-teller sentences, as are shown in the second and third provable conditionals about them.

Example 3. The Solovayesque sentences on $\mathcal{F}$, in which the domain is $\{0, 1, 2\}$, and the binary relation is $\{\langle 0, 1\rangle, \langle 0, 2\rangle\}$.

In this example, we have three Solovayesque sentences on $\mathcal{F}$, say σ_0, σ_1, and σ_2. We list the provable sentences about these three sentences as follows: $\sigma_0 \vee \sigma_1 \vee \sigma_2$, $\sigma_i \to \neg\sigma_j$ $(i, j = 0, 1, 2,\ i \neq j)$, $\sigma_0 \to T\ulcorner\sigma_1\urcorner$, $\sigma_0 \to T\ulcorner\sigma_2\urcorner$, $\sigma_1 \to \neg T\ulcorner\sigma_1\urcorner$, and $\sigma_2 \to \neg T\ulcorner\sigma_2\urcorner$. For convenience, these sentences will be called the 'baptising sentences' for σ_0, σ_1, and σ_2.

We come to show that $\sigma_1 \leftrightarrow \neg T\ulcorner\sigma_1\urcorner$ is provable. For this, first note that $\sigma_0 \to T\ulcorner\sigma_2\urcorner$, together with $\sigma_0 \to T\ulcorner\sigma_1\urcorner$, implies that $\sigma_0 \to T\ulcorner\sigma_1\urcorner \wedge T\ulcorner\sigma_2\urcorner$ is provable. Besides, by $\sigma_1 \to \neg T\ulcorner\sigma_1\urcorner$ and $\sigma_2 \to \neg T\ulcorner\sigma_2\urcorner$, we have $T\ulcorner\sigma_1\urcorner \wedge T\ulcorner\sigma_2\urcorner \to \neg\sigma_1 \wedge \neg\sigma_2$. Also, we know $\neg(\sigma_1 \vee \sigma_2) \to \sigma_0$ from $\sigma_0 \vee \sigma_1 \vee \sigma_2$. It follows that $T\ulcorner\sigma_1\urcorner \wedge T\ulcorner\sigma_2\urcorner \to \sigma_0$ is provable. Consequently, we obtain $\sigma_0 \leftrightarrow T\ulcorner\sigma_1\urcorner \wedge T\ulcorner\sigma_2\urcorner$ is provable.

What is more, $\sigma_0 \vee \sigma_1 \vee \sigma_2$ implies $\neg\sigma_1 \to \sigma_0 \vee \sigma_2$, which, by the conclusion we just get, implies that $(T\ulcorner\sigma_1\urcorner \wedge T\ulcorner\sigma_2\urcorner) \vee \neg T\ulcorner\sigma_2\urcorner$. Then, $\neg\sigma_1 \to T\ulcorner\sigma_1\urcorner$ is provable. Since $\sigma_1 \to \neg T\ulcorner\sigma_1\urcorner$ is one of the baptising sentences, we obtain that $\sigma_1 \leftrightarrow \neg T\ulcorner\sigma_1\urcorner$ is provable. Similarly, we can get that $\sigma_2 \leftrightarrow \neg T\ulcorner\sigma_2\urcorner$ is also provable.

We have seen that the sentences σ_1 and σ_2 are nothing but the liar sentence. Again, like the paradox in Example 1, the Soloveyesque paradox in Example 3 is really a known one, only it is more deeply hidden, so more difficult to uncover its disguise.

Example 4. The Solovayesque sentences on $\mathcal{F}$, in which the domain is $\{0, 1, 2\}$, and the binary relation is $\{\langle 0, 1\rangle, \langle 1, 2\rangle\}$.

This example is very similar to the previous one. The present baptising sentences are the same as these in the previous example except that we now have $\sigma_1 \to T\ulcorner\sigma_2\urcorner$ rather than $\sigma_0 \to T\ulcorner\sigma_2\urcorner$. For definiteness, we list the last four baptising sentences: $\sigma_0 \to T\ulcorner\sigma_1\urcorner$, $\sigma_1 \to T\ulcorner\sigma_2\urcorner$, $\sigma_1 \to \neg T\ulcorner\sigma_1\urcorner$, and $\sigma_2 \to \neg T\ulcorner\sigma_2\urcorner$. Note that the middle two can be combined into $\sigma_1 \to \neg T\ulcorner\sigma_1\urcorner \wedge T\ulcorner\sigma_2\urcorner$. Also, the converse is provable. This is because the first conditional and the last one imply $\neg T\ulcorner\sigma_1\urcorner \wedge T\ulcorner\sigma_2\urcorner \to \neg\sigma_0 \wedge \neg\sigma_2$. This, together with $\neg\sigma_0 \wedge \neg\sigma_2 \to \sigma_1$, entails $\neg T\ulcorner\sigma_1\urcorner \wedge T\ulcorner\sigma_2\urcorner \to \sigma_1$. We thus get $\sigma_1 \leftrightarrow \neg T\ulcorner\sigma_1\urcorner \wedge T\ulcorner\sigma_2\urcorner$ is provable.

There are still two baptising conditionals, say $\sigma_0 \to T\ulcorner\sigma_1\urcorner$ and $\sigma_2 \to \neg T\ulcorner\sigma_2\urcorner$, which we may wonder whether the converses of them are also provable. If so, just as we see in the previous examples, we would again obtain a known paradox with disguise. Interestingly, in this case, the two conditionals are the best things we can obtain about σ_0 and σ_1. We will prove this observation in the next section.

For now, let us just strengthen one of the above two baptising conditionals. First, we can add the converse of $\sigma_0 \to T\ulcorner\sigma_1\urcorner$. Then, we get $\sigma_0 \leftrightarrow T\ulcorner\sigma_1\urcorner$ is provable. About σ_2, since $\neg\sigma_2 \leftrightarrow \sigma_0 \vee \sigma_1$, we can see $\sigma_2 \leftrightarrow \neg T\ulcorner\sigma_1\urcorner \wedge T\ulcorner\sigma_2\urcorner$. To sum up, the first strengthening is the sentences σ_0, σ_1, and σ_2 such that the following biconditionals are provable:

$$\begin{cases} \sigma_0 \leftrightarrow T\ulcorner\sigma_1\urcorner \\ \sigma_1 \leftrightarrow \neg T\ulcorner\sigma_1\urcorner \wedge T\ulcorner\sigma_2\urcorner \\ \sigma_2 \leftrightarrow \neg T\ulcorner\sigma_1\urcorner \wedge \neg T\ulcorner\sigma_2\urcorner \end{cases} \tag{3}$$

The alternative choice is to strengthen $\sigma_2 \to \neg T\ulcorner\sigma_2\urcorner$. In this case, we get the following provable biconditionals:

$$\begin{cases} \sigma_0 \leftrightarrow T\ulcorner\sigma_1\urcorner \wedge T\ulcorner\sigma_2\urcorner \\ \sigma_1 \leftrightarrow \neg T\ulcorner\sigma_1\urcorner \wedge T\ulcorner\sigma_2\urcorner \\ \sigma_2 \leftrightarrow \neg T\ulcorner\sigma_2\urcorner \end{cases} \tag{4}$$

Now, it should be noted that like these we give in Example 1 and 2, the sentences σ_0, σ_1, and σ_2 satisfying (3) or (4) are are 'traditional' or 'ordinary' in that any of them are of form '$\sigma \leftrightarrow \ldots T\ulcorner\sigma'\urcorner \ldots$', in which '$\ldots T\ulcorner\sigma'\urcorner \ldots$' is a sentence containing at least a sentence of form $T\ulcorner\sigma'\urcorner$. Intuitively, these sentences always make a straightforward assertion about the truth or falsity of some sentences. The provable biconditionals that we use to characterise the ordinary paradoxical sentences are nothing but the formalisation of these assertions.

By contrast, we must have noticed that the Solovayesque sentences are those asserting some arithmetic function converges to a number. They never straightforwardly say of the truth or falsity of any sentence. We do get some provable conditionals that relate these sentences to those saying of truth or falsity of them. However, these conditionals are secondary, and they are unlikely to provide complete information about the above relation. We can anticipate that there is a situation in which the converse of some of these conditionals fails. It really comes true for the sentences we just construct in Example 4. We will discuss this topic in more detail in the next section.

5 Comparison

The Solovayesque sentences that we construct on, by Theorem 1, are paradoxical, if the frame embedded into these sentences has at least two points, and contains only the directed cycle of length 1. We give some typical Solovayesque sentences in the previous section. Some of them are just well-known paradoxes or pathological sentences (Example 1 and 3). More importantly, others are radically different from all the know ones in that the provable sentences that are used to characterise them are not all biconditionals (Example 4). In this section, we will give a formal proof of this point.

To see that the baptising sentences in Example 4 are not logically equivalent to those in (3) or (4), we only need to construct a model in which the baptising sentences in Example 4 are all true, while the sentence $T\ulcorner\sigma_1\urcorner \rightarrow \sigma_0$ ($\neg T\ulcorner\sigma_2\urcorner \rightarrow \sigma_2$, respectively) is false. For $T\ulcorner\sigma_1\urcorner \rightarrow \sigma_0$, just let X, that is, the extension of T, be $\{\sigma_1\}$. Notice that sentence σ_0, σ_1, and σ_2 are something like variables except that they are subject to their baptising sentences. Thus, we can almost freely fix their truth values except that we must consider the truth conditions for their baptising sentences. A suitable choice is such that $X \models \neg\sigma_0 \wedge \neg\sigma_1 \wedge \sigma_2$. In this way, we can see that in the model $\langle\mathfrak{N}, X\rangle$, the baptising sentences in Example 4 are all true, but $T\ulcorner\sigma_1\urcorner \rightarrow \sigma_0$ is false. Similarly, by choose $X = \varnothing$ and $X \models \sigma_0 \wedge \neg\sigma_1 \wedge \neg\sigma_2$, we can obtain that $\neg T\ulcorner\sigma_2\urcorner \rightarrow \sigma_2$ is not a logical consequence of the baptising sentences in Example 4.

In the above, we have proved that the baptising sentences in Example 4 are properly weaker than the sentences in (3) or (4). We can see that the former may be seen as obtained from the latter by weakening some baptising sentences. This is essentially a syntactic difference. It has nothing to do with the truth predicate. We will give a possible-world semantics, in which T would be something like the truth predicate: $T\ulcorner A\urcorner$ and A are equivalent across possible worlds. Under the interpretation, both the set of the baptising sentences in Example 4 and the set of sentences in (3) or (4) are still paradoxical. Above all, the former has a lower degree of paradoxical than the latter in the sense we will explain below.

Definition 2. *A valuation in a frame $\mathcal{F} = \langle W, R\rangle$ is a mapping from W to the power set of $\mathbb{N}$. A valuation V in $\mathcal{F}$ is admissible for a sentence A, if for any points $u, v \in W$ with $u\,R\,v$,*

$$V(v) \models T\ulcorner A\urcorner \iff V(u) \models A. \tag{5}$$

The above notion, from [3, pp. 243-244], gives a new interpretation for the truth predicate symbol T. Informally, equation (5) says that the sentence that asserts the truth of A is true (false) at a point v, then A itself is true (false) at any accessible point from v. Thus, although $T\ulcorner A\urcorner$ and A are not logically equivalent, they are equivalent with respect to two accessible worlds. Briefly, they are equivalent across possible worlds. See [3] for more details. For now, we make it clear how to distinguish the paradoxes by this definition.

The following notion is also from [3].

Definition 3. *A set of sentences, say Σ, is paradoxical in $\mathcal{F}$, if no valuation in $\mathcal{F}$ is admissible for (all sentences of) Σ. In that case, $\mathcal{F}$ is called a characterisation frame for Σ.*

Note that Definition 2 is collapsed to Definition 1 when $\mathcal{F}$ is the minimal reflexive frame. Thus, the minimal reflexive frame is a characterisation frame for any paradoxical set. However, we pay more attention to the difference between paradoxes concerning their characterisation frames. This is the way we distinguish a paradox from another one. For instance, the characterisation frames for the liar paradox are properly included in those for the Jourdain's card paradox (i.e., the sentences δ_0 and δ_1 such that $\delta_0 \leftrightarrow \neg T\ulcorner\delta_1\urcorner$ and $\delta_1 \leftrightarrow T\ulcorner\delta_0\urcorner$ are provable). See [3, pp. 254-255] for the proof.

We now consider the paradox that we give in Example 4. Recall that it consists of three sentence σ_0, σ_1, and σ_2 such that $\sigma_0 \vee \sigma_1 \vee \sigma_2$, $\sigma_i \to \neg\sigma_j$ $(i, j = 0, 1, 2,\ i \neq j)$, $\sigma_0 \to T\ulcorner\sigma_1\urcorner$, $\sigma_1 \to T\ulcorner\sigma_2\urcorner$, $\sigma_1 \to \neg T\ulcorner\sigma_1\urcorner$, and $\sigma_2 \to \neg T\ulcorner\sigma_2\urcorner$ are all provable. We use Σ for the set of these sentences σ_0, σ_1, and σ_2. We will compare it with the set of sentences satisfying equation (3). We use Σ' for the latter. Let us consider the frame in Figure 1, in which the domain $W = \{0, 1\}$, and the binary relation $R = \{\langle 0, 1\rangle, \langle 1, 0\rangle\}$.

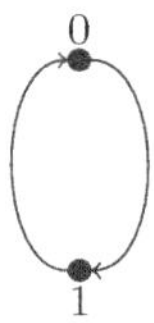

$T\ulcorner\sigma_0\urcorner$	$T\ulcorner\sigma_1\urcorner$	$T\ulcorner\sigma_2\urcorner$	σ_0	σ_1	σ_2
T	F	T	F	T	F
F	T	F	F	F	T

Fig. 1. An admissible valuation for Example 4

Claim (1). The set $\{\sigma_0, \sigma_1, \sigma_2\}$ is not paradoxical in the frame in Figure 1.

For brevity, we use Σ for the set $\{\sigma_0, \sigma_1, \sigma_2\}$, and $\mathcal{F}$ for the frame in question. To prove Σ is not paradoxical in $\mathcal{F}$, it suffices to find a valuation in $\mathcal{F}$, which is admissible for Σ. Fix a valuation V in $\mathcal{F}$ as follows: $V(0) = \{\sigma_0, \sigma_2\}$ (more precisely, the set of the Gödel numbers of σ_0, σ_2), and $V(1) = \{\sigma_1\}$. Thus, we have $V(0) \models T\ulcorner\sigma_0\urcorner$, $V(0) \models T\ulcorner\sigma_2\urcorner$, but $V(0) \not\models T\ulcorner\sigma_1\urcorner$. At point 1, we have $V(1) \models T\ulcorner\sigma_1\urcorner$, but neither $V(1) \models T\ulcorner\sigma_0\urcorner$ nor $V(1) \models T\ulcorner\sigma_2\urcorner$. Thus, we have known the truth values of all the sentences $T\ulcorner\sigma_i\urcorner$ $(i = 0, 1, 2)$ at any point of $\mathcal{F}$. By the way, whenever $V(u) \models A$, we will say A is true at point u (under the valuation V).

By the soundness for the first-order logic, any provable sentence (in PAT) must be true for any extension of T, that is, $\langle \mathfrak{N}, X\rangle \models A$ holds for any provable A. Therefore, the sentences that baptise σ_0, σ_1, and σ_2 must be true at any point of $\mathcal{F}$. Such a requirement is useful to deduce the truth values of σ_0, σ_1, and σ_2. For instance, from $V(0) \models \sigma_1 \to \neg T\ulcorner\sigma_1\urcorner$, together with $V(0) \models T\ulcorner\sigma_1\urcorner$, we can deduce $V(0) \not\models \sigma_1$, that is to say, σ_1 must be false at point 0. At the same time, we must realise that the baptising sentences for Σ do not provide enough information to determine the truth values of all σ_0, σ_1, and σ_2 at any point of $\mathcal{F}$. As we will see below, this is quite different from the case of the paradox that serves as a contrast.

We aim to prove that $\mathcal{V}$ is admissible for Σ. The current problem is that we still need to more information about $\mathcal{V}$. For this purpose, we can determine the truth values of all σ_0, σ_1, and σ_2 by supposing that $\mathcal{V}$ is admissible for Σ. In this way, we at least make V satisfy the admissibility condition. After doing that, it will suffice for us to verify that all the baptising sentences for Σ are valid in $\mathcal{F}$ (viz., true at any point in $\mathcal{F}$).

By equation (5), we can deduce from $V(0) \not\models T\ulcorner\sigma_0\urcorner$, together with $0R1$, that $V(1) \not\models \sigma_0$, that is, σ_0 is false at point 1. Similarly, we can get that at point 1, σ_1 is false, and σ_2 is true. At point 0, only σ_2 is true. All of these truth values are shown on the right side of Figure 1. It is clear that the valuation V, if holds these truth values for Σ in addition, must be admissible for Σ.

It remains to verify that the baptising sentences for Σ are valid in $\mathcal{F}$. For instance, since σ_2 is true at point 0 and σ_1 is true at point 1, $\sigma_0 \vee \sigma_1 \vee \sigma_2$ is valid in $\mathcal{F}$. Also, since σ_0 is false at point 0 and 1, we know $\sigma_0 \rightarrow T\ulcorner\sigma_1\urcorner$ is valid in $\mathcal{F}$. Similarly, we can verify all the other baptising sentences for Σ are also valid in $\mathcal{F}$. We leave the details to the reader.

It should be noted that $T\ulcorner\sigma_1\urcorner \rightarrow \sigma_0$ is false at point 1 in $\mathcal{F}$. Therefore, among the baptising sentences for Σ, there is at least one which cannot any longer be strengthened to be the biconditional $\sigma_0 \leftrightarrow T\ulcorner\sigma_1\urcorner$. We now give a claim to show the set Σ is different from the Σ'.

Claim (2). The set Σ' is paradoxical in the frame in Figure 1.

Note that Σ' belongs to the so-called Boolean paradoxes. For this kind of paradoxes, we have known how to determine all of their characterisation frames. From this, a proof of the above claim is immediate. For more details, we refer the reader to [4, Theorem 1.7, p. 885]. For the sake of the self-containedness, we here give a straightforward proof.

An important feature of Σ' is that the baptising sentences for Σ' are all biconditionals. We will make full use of this feature to deduce that no valuation in $\mathcal{F}$ is admissible for Σ'. Assume V is admissible for Σ'. For convenience, for $i = 0, 1$, we use $i^- = 1, 0$. Then, by the admissibility of $\mathcal{V}$, together with the validity of the baptising sentences for Σ' in $\mathcal{F}$, we can obtain the following equivalences:

(i) $V(i) \models \sigma_0$, iff $V(i^-) \models \sigma_1$.
(ii) $V(i) \models \sigma_1$, iff $V(i^-) \not\models \sigma_1$ and $V(i^-) \models \sigma_2$.
(iii) $V(i) \models \sigma_2$, iff $V(i^-) \not\models \sigma_1$ and $V(i^-) \not\models \sigma_2$.

Case 1: suppose $V(0) \models \sigma_0$, then by (i), $V(1) \models \sigma_1$. On the other hand, by (ii), $V(0) \models \sigma_2$, and by (iii), it follows $V(1) \not\models \sigma_1$, a contradiction!

Case 2: suppose $V(0) \not\models \sigma_0$, then by (i), $V(1) \not\models \sigma_1$. By (ii), we have either $V(0) \models \sigma_1$ or $V(0) \not\models \sigma_2$.

Case 2.1: suppose $V(0) \models \sigma_1$, then the left-to-right direction of (ii) implies $V(1) \models \sigma_2$. At the same time, the right-to-left direction of (iii) tells us $V(1) \not\models \sigma_2$, a contradiction!

Case 2.1: suppose $V(0) \not\models \sigma_2$, then by case 2.1, we can suppose further $V(0) \not\models \sigma_1$. Now, by (iii), we have $V(1) \models \sigma_2$. Since $V(1) \not\models \sigma_1$, by (ii), we get $V(0) \models \sigma_1$, a contradiction!

To sum up, we can conclude that Σ' is paradoxical in $\mathcal{F}$.

Definition 4. *Σ_1 has a (strictly) higher degree of paradoxicality than Σ_2, if the characterisation frames for Σ_1 (properly) include those for Σ_2.*[4]

We notice that any sentence in Σ is a logical consequence of Σ'. Hence, if a valuation is admissible for Σ', it must be so for Σ. The following result is a summary of this observation and the two claims we just prove above: the characterisation frames for Σ are properly included in those for Σ'.

We should point out that the above result also holds when Σ' is the set satisfying (4). For this, a key observation is that the valuation in the frame shown in Figure 2 is admissible for Σ, while the present Σ' is paradoxical in this frame. We leave the proof to the reader.

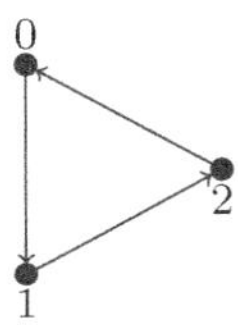

$T\ulcorner\sigma_0\urcorner$	$T\ulcorner\sigma_1\urcorner$	$T\ulcorner\sigma_2\urcorner$	σ_0	σ_1	σ_2
F	F	T	F	T	F
F	T	F	F	F	T
F	T	F	F	T	F

Fig. 2. Another admissible valuation for Example 4

In conclusion, we obtain the following result:

Proposition 3. *The Solovayesque paradox in Example 4 has a strictly lower degree of paradoxicality than the paradox in equation (3) or (4).*

As mentioned in Section 4, equation (3) and (4) are two possibilities that we can strengthen the Solovayesque sentences in Example 4. We have proved that both the paradoxes corresponding to equation (3) and (4) have a higher degree of paradoxicality than the Solovayesque paradox in Example 4. It is evident that the first two paradoxes are ordinary in that their baptising sentences are all bicondtionals. From Proposition 3, We can see that at least one baptising sentences for the Solovayesque paradox in Example 4 cannot be strengthened to be biconditional. From this perspective, the Solovayesque paradox in Example 4 is indeed radically different from the two ones that we use as a contrast. To some degree, we can take the Solovayesque paradox in Example 4 as the paradox obtained from the ordinary paradoxes by weakening some of these baptising biconditionals. It is a paradox with broken symmetry, while the ordinary paradoxes, including the two in equation (3) and (4), all have the perfect symmetry.

6 Conclusion

Our construction is derived from Solovay's construction in his proof of the arithmetical completeness of Gödel-Löb provability logic. The core is still a diago-

[4] See [3, p. 254] or [4, p. 885].

nalisation of the so-called Solovay function. However, our construction employs the truth predicate instead of the provability predicate in the formalisation of the Solovay function. The sentences we construct, as is shown in Theorem 1, are paradoxical, provided the finite frame satisfies some conditions. It is well-known that the sentences in Solovay's construction are all undecidable. The paradoxical sentences we get are in the same vein as the original Solovay sentences, and thus may be seen as the paradoxes underlying the latter.

One highlight of our construction is that some of the paradoxes that we construct are beyond the scope of all the known paradoxes. As we have pointed out before, some paradoxes in our construction are merely those well-known ones, such as the liar paradox (Example 1, 3). Nonetheless, our construction also provides some paradoxes that we did not yet realise before. Like all the known paradoxical (and other pathological) sentences, the Solovayesque sentences are also formalised by the diagonal method. Also, they are determined by the fixed points of some formulas. However, unlike the former, the latter cannot be completely characterised by the biconditionals. In other words, among all the baptising sentences for the Solovayesque sentences, at least one of them is not a biconditional.

In Section 5, we prove that the Solovayesque sentences in Example 4 are the typical ones of this kind. It may be seen as a paradox of some broken symmetry relative to the two ordinary paradoxes determined respectively by equation (3) and (4). Indeed, as a paradox, it is even weaker than the two ordinary ones. At this point, we prove that the former has a strictly lower degree of paradoxcality than the latter two.

There are still some issues worth our further study. For instance, we notice that among the Solovayesque paradoxes, some are just the known ones, others are the ones we highlight in this paper—those with the broken symmetry. The difference, of course, lies in the finite frame on which we construct the Solovayesque sentences. A natural question is on what frames we can construct the Solovayesque paradoxes that have the broken symmetry. Also, so far, we can only attribute the broken symmetry to the form of their baptising sentences. Another question is whether such a feature can be described by use of the characterisation frames. It may be possible to articulate the symmetry of the paradoxes by some symmetry of their characterisation frames.

About Solovay's arithmetical completeness theorem, Smoryński, in his book [5, p. 148], commented that Solovay's original construction is 'a powerful tool in obtaining refined incompleteness results'. In particular, we can construct the undecidable sentences 'illustrating the types of incompleteness phenomenon desired' by choosing Kripke models. For instance, the 'extremely undecidable sentences' that Boolos [2] gave are quite different from Gödel's sentence and Rosser's sentences. From this point of view, the construction we give in this paper is a tool to generate refined paradoxicality. On Kripke frames, we can construct the various kinds of paradoxical sentences, which include not only the known ones such as the liar and Jourdain's card paradox but also those that we still need to explore further.

References

1. Boolos, G.: Logic of Provability. Cambridge University Press, Cambridge (1993)
2. Boolos, G.: Extremely undecidable sentences. The Journal of Symbolic Logic **47**(1), 191–196 (1982)
3. Hsiung, M.: Jump Liars and Jourdain's Card via the relativized T-scheme. Studia Logica **91**(2), 239–271 (2009)
4. Hsiung, M.: Boolean paradoxes and revision periods. Studia Logica **105**(5), 881–914 (2017)
5. Smoryński, C.: Self-Reference and Modal Logic. Springer-Verlag, New York (1985)
6. Solovay, R.M.: Provability interpretations of modal logic. Israel Journal of Mathematics **25**(3), 287–304 (1976)

Multiple Roles and Deontic Logic*

Ziyue Hu

Renmin University of China
huziyue@ruc.edu.cn

Abstract. This paper aims to introduce the concept of role to characterize the tolerance for normative conflicts. It starts by presenting a kind of syllogism, which we call the role-related deontic syllogism. Then, we propose a deontic logic of multiple roles based on the role-related deontic syllogism. The significant characteristic of this logic is tolerating normative conflicts. A sound and complete deductive system for this logic is presented, and the decidability of this logic is proved. In addition, based on this logic, we discuss two different modes of collective obligation.

Keywords: Multiple Roles · Deontic Logic · Normative Conflict· Collective Obligation

1 Introduction

The tolerance for normative conflicts is an attractive topic of discussion in deontic logic, which is mainly because standard deontic logic (SDL) is not tolerant of conflicting obligations. SDL is the weakest normal modal logic of type KD in the Chellas classification [2]. The conjunction $O\varphi \wedge O\neg\varphi$ is the general representation of the normative conflict, which means two contradictory propositions are simultaneously obligatory. Since D: $O\varphi \rightarrow \neg O\neg\varphi$ is an axiom of SDL and $O\varphi \rightarrow \neg O\neg\varphi$ is semantically equivalent to $\neg(O\varphi \wedge O\neg\varphi)$, normative conflicts are ruled out by SDL, which is supported by a rationalist perspective (cf. [3]). However, in our daily life, the occurrence of normative conflicts is common and is not necessarily indicative of a logical contradiction.

Some non-normal modal logics are capable of tolerating normative conflicts. A typical example is the non-normal modal logic EM. Neighborhood semantics [7] is a conventional semantics purposed to interpret the non-normal modal logic. In addition, multi-relational semantics [1] is applicable to interpret the non-normal logic, and a kind of preference semantics that is closely associated with multi-relational semantics [4, 5] is developed to characterize the tolerance for normative conflicts.

In this paper, we introduce the concept of role to analyze the tolerance for normative conflicts and provide a semantics that is different from those mentioned above. Although the difference between roles sometimes leads to a normative conflict, individuals with multiple roles still accept obligations according

* This chapter is in its final form and it is not submitted to publication anywhere else.

B. Liao and Y. N. Wáng (eds.), *Context, Conflict and Reasoning*,
Logic in Asia: Studia Logica Library,
https://doi.org/10.1007/978-981-15-7134-3_3

to different perspectives of diverse roles in practice. When a normative conflict is confronted by an individual assigned with multiple roles, it does not suggest that this individual makes a logical mistake in thinking. By treating roles as a tool for constructing organizations, some authors apply the concept of roles to specify the delegation under the context of an organization(cf. e.g., [9, 6]). In this article, the roles under consideration generally refer to those that are often mentioned in daily life to show the particularity of personal identities, such as father, teacher, and doctor. The story of Sartre's student in [8] is representative of the normative conflict caused by multiple roles. As a patriot and revenger, the student is obligated to leave home for joining the Free France Forces. Nevertheless, as a son with the responsibility to care for his mother, the student is duty-bound to stay home. Sartre takes the view that there is nothing right or wrong regarding these two options.

The rest of this paper is organized as follows. Section 2 introduces the role-related deontic syllogism and some relevant symbolic expressions. In Section 3, a deontic logic of multiple roles is proposed. In Section 4, we give an axiomatic system and prove the corresponding soundness and completeness theorems. In Section 5, decidability is also proved. In Section 6, a discussion is conducted about two modes of collective obligation. The conclusion of this paper and future works are presented in Section 7.

2 Role-Related Deontic Syllogism

Typically, individuals are assigned multiple roles in their social life, which results from a variety of internal or external factors, for example, social evaluation, cultural background, personality, occupation, and educational level. We can adopt an analytical perspective on various roles, that is to say, regarding them as different combinations of influences from these internal or external factors. Such influence imposes various normative constraints on different roles. Therefore, the concept of role is crucial for understanding the rationality of normative constraints in social life. Intrinsically, it is inevitable that all normative statements depend on the specific roles for indicating the rationality of their normative constraints, such as "maintaining public order is obligatory for the police officer," and "the administrator with the highest authority ought to close all redundant programs." It is easy to find out that, in the practice of language, using a subject representing a particular role in a normative statement is conducive to ensuring the rationality of some normative constraints, which is mainly because a specific role is capable of showing the rationality of some normative constraints by evoking or associating some acceptable scenarios similar to the actual scenario. Moreover, people can be distinguished through various roles, which makes each role correspond to a particular group.

We call those obligations dependent on a role the role-related obligations. Corresponding to the role-related obligations, the obligations placed on an individual can be referred to as the individual obligations. By introducing an operator O_s that represents the concept of the individual obligations placed on Sartre's student and using φ to represent staying home, the expression $O_s\varphi \wedge O_s\neg\varphi$ can be used to represent the normative conflict that Sartre's student encounters. When an individual with multiple roles considers what his/her obligations are, it is natural that those role-related obligations corresponding to his/her roles will be considered through a reasoning process from the role-related obligation to the individual obligation. We believe this reasoning process takes place in a syllogistic form. An ostensibly acceptable version of this syllogistic form can be reflected by the following example:

Major Premise:	Maintaining public order is obligatory for police officers.
Minor Premise:	Agent α is a police officer.
Conclusion:	Maintaining public order is obligatory for agent α.

Table 1.

Based on the above syllogistic argument, it can be judged that the individual obligations to be fulfilled by an agent are determined by what roles the agent holds. Thus, if two distinct roles assigned to an agent bring a couple of conflicting role-related obligations, then the agent's individual obligations will eventually lead to a normative conflict. In daily life, this phenomenon is not uncommon since people encounter normative conflicts often due to their multiple roles.

However, the above syllogistic argument lays more emphasis on the theoretical level to derive the individual obligation by the agent's roles, while it ignores the practical limitation of roles. If an individual considers his/her individual obligations according to the pattern of the above syllogistic argument, he/she will be bound by the role-related obligations that are placed on his/her roles in all practical scenarios. Therefore, it is inevitable to answer whether the above syllogistic argument is applicable in all practical scenarios.

In practice, it is frequently impractical to take into consideration our individual obligations based on all our roles, which dues to the factor that some sensible or practical restrictions tend to make individuals choose only one or several roles to consider their obligations and ignore the remaining roles. More succinctly, people only choose their reasonable roles in practice. For example, if a firefighter on leave takes his daughter to a theater for watching a Disney movie and get caught in a fire, the firefighter would consider his obligations based on his role as a father rather than his role as a fireman. As for Sartre's student's story, we can think that two reasonable roles lead to a normative conflict.

As discussed above, it is inappropriate to apply the aforementioned syllogistic argument in all practical scenarios, because this syllogistic argument cannot ensure the role of a police officer is reasonable for agent α in all practical scenarios. However, the aforementioned syllogistic argument means agent α should invariably bear a police officer's obligations in all practical scenarios if agent α is a police officer. Hence, such a reasoning process is unacceptable.

Therefore, we believe the minor premise in the aforementioned syllogistic argument is inappropriate to serve as the bridge from the individual obligation to the individual obligation. Because it only represents the role eligibility but ignores the factor of practical limitations of roles. With this idea in mind, we think that the most effective solution is to present a more appropriate statement form to serve as the minor premise that can restrict the practical scope of roles. We adopt the statement form "agent α reasonably acts as role r" to serve as the appropriate statement form of the minor premise. By this form, we can refine the syllogistic argument shown in Table 1.

Major Premise:	Maintaining public order is obligatory for police officers.
Minor Premise:	Agent α reasonably acts as a police officer.
Conclusion:	Maintaining public order is obligatory for agent α.

Table 2.

Based on the syllogistic pattern shown in Table 2, a kind of syllogism is obtained, which we call the role-related deontic syllogism. If the new minor premise is true, which means this premise is a fact, we can realize that there are reasons in the practical scenario to support agent α to consider the individual obligations according to the role of a police officer without violating practical rationality. Therefore, this new minor premise indicates that the role of a police officer is reasonable for agent α in practice. Conversely, if the role of a police officer is reasonable for agent α in practice, it is not arguable that agent α reasonably acts as a police officer. Besides, from this new minor premise, it can be known that agent α is one member of police officers, and agent α is qualified for the role of a police officer.

We use the expression $AS_\alpha(r)$ to represent "agent α reasonably acts as role r", use the expression $O^r\varphi$ to represent "φ is obligatory for role r" and use $O_\alpha\varphi$ to represent "φ is obligatory for agent α". In this way, the form of the role-related deontic syllogism can be formalized as the following expression:

$$O^r\varphi \wedge AS_\alpha(r) \rightarrow O_\alpha\varphi.$$

The expression $O^r\bot$ means that the logical falsehood can be derived from those propositions that are obligatory for role r. Intuitively, it is unacceptable if role r is reasonable for agent α in practice. Besides, if we accept the expression $O^r(\varphi \rightarrow \psi) \rightarrow (O^r\varphi \rightarrow O^r\psi)$ and the necessitation rule, then any proposition is obligatory for role r when $O^r\bot$ is true. It is evident that we need to avoid this ridiculous situation. Therefore, if agent α reasonably acts as role r, then it is false that the logical falsehood is obligatory for role r. Thus, we can get the following expression:

$$AS_\alpha(r) \rightarrow \neg O^r\bot.$$

Since both capability and life are restricted for everyone, the number of roles that an individual can hold is limited. Let *Role* be a finite set of different roles. If we accept that all individual obligations stem from the role-related obligations, or adopt an attitude that considering the individual obligations only through various roles, the expression $O_\alpha\varphi$ can be introduced by the following expression:

$$O_\alpha\varphi =_{df} \bigvee_{i\in Role} (O^i\varphi \wedge AS_\alpha(i)).$$

According to the above expression, any two conflicting individual obligations can be attributed to a conflict of role-related obligations. More crucially, agent α can be tolerant of normative conflicts by the above expression. Besides, due to the expression $AS_\alpha(i) \rightarrow \neg O^i\bot$, the logical falsehood cannot be obligatory for agent α. Therefore, we can get the following expression:

$$\neg O_\alpha\bot.$$

If only emphasizing the tolerance for normative conflicts, it will result in an undesirable situation, that is, all propositions and their negations are simultaneously obligatory. In reality, there are some norms that must be accepted unconditionally in any practical scenario, even if an individual takes on some particular roles. Therefore, regardless of what roles an agent holds, there is a class of obligations that must be accepted, which is called the all-things-considered obligation. With this kind of obligation, we can avoid the above-mentioned undesirable situation. The correlation between the role-related obligation and the all-things-considered obligation needs to abide by the following expression:

$$O\varphi \rightarrow O^i\varphi.$$

Furthermore, the all-things-considered obligation should meet the following condition:

$$O\varphi \rightarrow \neg O\neg\varphi.$$

Then, based on the above two expressions and the definition of $O_\alpha\varphi$, the correlation between the individual obligation and the all-things-considered obligation should be constrained by the following condition:

$$O_\alpha\varphi \rightarrow \neg O\neg\varphi.$$

This expression means the individual obligation cannot conflict with the all-things-considered obligation, which can be used to explain why the all-things-considered obligation can prevent the above-mentioned undesirable situation.

3 A Deontic Logic of Multiple Roles

In this section, we give a deontic logic of multiple roles(DL-MR). The primary idea of DL-MR is to characterize the agent's individual obligations based on multiple roles.

3.1 Syntax and Semantics

Definition 1. *Assume a countable set Atm of atomic propositions, a finite set Agt of agents, and a finite set Role of roles. The language $\mathcal{L}$ is defined by the following grammar:*

$$\varphi ::= p \mid \neg\varphi \mid \varphi \rightarrow \psi \mid O\phi \mid O^i\varphi \mid AS_\alpha(i) \mid O_\alpha\varphi \mid \bot$$

where $p \in Atm, i \in Role$ and $\alpha \in Agt$.

The intended reading of $O\varphi$ is "φ is obligatory". Formulas of the form $O^i\varphi$ and $O_\alpha\varphi$ should be read, respectively, as "φ is obligatory for role i" and "φ is obligatory for agent α". Formulas of the form $AS_\alpha(i)$ express that "agent α reasonably acts as role i". We can regard formulas of the form $AS_\alpha(i)$ as particular atomic propositions that are different from the atomic propositions in *Atm*. The propositional connectives $(\wedge, \vee, \leftrightarrow)$ and the propositional constant $\top$ are defined in the standard way. Some deontic operators can be defined via abbreviations as usual, i.e., $P := \neg O \neg$, $P^i := \neg O^i \neg$, $F := O\neg$, $F^i := O^i\neg$.

Definition 2. *A **MR**-model is a tuple $\mathcal{M} = \langle W, R, (R_i)_{i\in Role}, C, V\rangle$ where:*

- *W is a set of possible ideal worlds;*
- *R is a serial relation on W;*
- *each R_i is a subset of R;*
- *C (called the role function) is a function $C : W \times Agt \rightarrow \mathcal{P}(Role)$ such that for any pair $(w, \alpha) \in W \times Agt$, $C(w, \alpha) \subseteq \{i \in Role : R_i(w) \neq \emptyset\}$;*
- *$V : Atm \rightarrow 2^W$ is a valuation function.*

We use $R(w)$ to express the set of R–*successors* of w, i.e., $\{w' \in W : Rww'\}$. Similarly, for any role $i \in Role$, let $R_i(w)$ denote the set of R_i-*successors* of w.We use **MR** to denote the class of all **MR**-models.

Definition 3. *Given a **MR**-model $\mathcal{M} = \langle W, R, (R_i)_{i\in Role}, C, V\rangle$, a possible ideal world $w \in W$ and a formula $\varphi \in \mathcal{L}$, we can define the satisfaction relation $\models$ by induction on φ:*

$\mathcal{M}, w \models p$	iff	$w \in V(p)$
$\mathcal{M}, w \models AS_\alpha(i)$	iff	$i \in C(w, \alpha)$
$\mathcal{M}, w \models \neg\varphi$	iff	not $\mathcal{M}, w \models \varphi$
$\mathcal{M}, w \models \varphi \rightarrow \psi$	iff	$\mathcal{M}, w \not\models \varphi$ or $\mathcal{M}, w \models \psi$
$\mathcal{M}, w \models O\varphi$	iff	for any $w' \in R(w)$, $\mathcal{M}, w' \models \varphi$
$\mathcal{M}, w \models O^i\varphi$	iff	for any $w' \in R_i(w)$, $\mathcal{M}, w' \models \varphi$
$\mathcal{M}, w \models O_\alpha\varphi$	iff	there exists a $r \in Role$ s.t. $\mathcal{M}, w \models O^r\varphi$ and $\mathcal{M}, w \models AS_\alpha(r)$
$\mathcal{M}, w \models \bot$		never

The truth set of φ is the set $[\![\varphi]\!]_\mathcal{M} = \{w \in W : \mathcal{M}, w \models \varphi\}$. If the context is clear, the subscript $\mathcal{M}$ can be omitted.

Definition 4. *Given a* ***MR****-model* $\mathcal{M} = \langle W, R, (R_i)_{i \in Lab}, C, V \rangle$ *and a formula* $\varphi \in \mathcal{L}$. φ *is satisfiable in* $\mathcal{M}$ *if there is some* $w \in W$ *such that* $\mathcal{M}, w \models \varphi$; φ *is globally satisfied in* $\mathcal{M}$ *(notation:* $\mathcal{M} \models_{\mathbf{MR}} \varphi$*) if for all* $w \in W$, $\mathcal{M}, w \models \varphi$; φ *is satisfiable if* φ *is satisfiable in a model in* ***MR****;* φ *is valid (notation:* $\models_{\mathbf{MR}} \varphi$*) if* φ *is globally satisfied in all models in* ***MR****. We use* ***DL-MR*** *to denote the set of all valid formulas.*

3.2 Some Logical Properties

In this part, we mention some logical properties of the defined operators.

Proposition 1. *For any* $\alpha \in Agt$, $\models_{\mathbf{MR}} O_\alpha \varphi \leftrightarrow \bigvee_{i \in Role} (O^i \varphi \wedge AS_\alpha(i))$.

Proof. According to the satisfaction condition for $O_\alpha \varphi$, $O_\alpha \varphi$ is semantically equivalent to $\bigvee_{i \in Role} (O^i_\alpha \varphi \wedge AS_\alpha(i))$. □

Proposition 2. *For any* $\alpha \in Agt$ *and any* $i \in Role$, $\not\models_{\mathbf{MR}} AS_\alpha(i) \leftrightarrow \neg O^i \bot$

Proof. We prove and explain this proposition by comparing the below three different **MR**-models in Figure 1. In $\mathcal{M}_1$, $R_i(w) \neq \emptyset$ and $i \in C(w, \alpha)$. Hence, $\mathcal{M}_1, w \models \neg O^i \bot$ and $\mathcal{M}_1, w \models AS_\alpha(i)$. It follows that $\mathcal{M}_1, w \models AS_\alpha(i) \leftrightarrow \neg O^i \bot$. In $\mathcal{M}_2$, since $i \notin C(w, \alpha)$ and $R_i(w) = \emptyset$, we have $\mathcal{M}_2, w \models \neg AS_\alpha(i)$ and $\mathcal{M}_2, w \models O^i \bot$. So $\mathcal{M}_2, w \models AS_\alpha(i) \leftrightarrow \neg O^i \bot$. In $\mathcal{M}_3$, since $i \notin C(w, \alpha)$ and $R_i(w) \neq \emptyset$, we have $\mathcal{M}_3, w \not\models AS_\alpha(i)$ and $\mathcal{M}_3, w \models \neg O^i \bot$. It follows that $\mathcal{M}_3, w \not\models AS_\alpha(i) \leftrightarrow \neg O^i \bot$. □

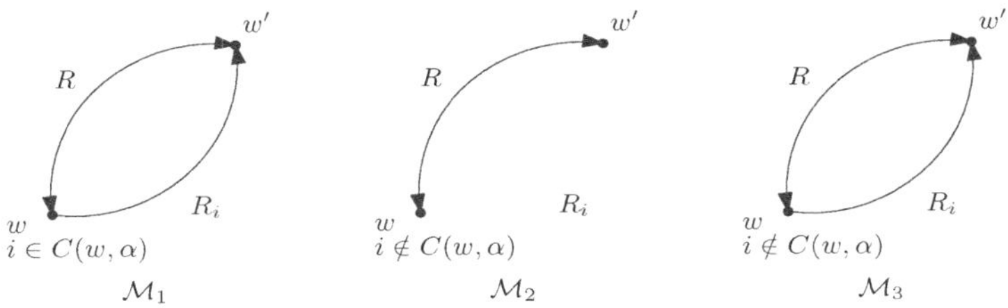

Fig. 1.

For any $\alpha \in Agt$, we can introduce two operators, one for individual permission, the other for individual prohibition:

- $P_\alpha \varphi =_{df} \bigvee_{i \in Role} (P^i \varphi \wedge AS_\alpha(i))$
- $F_\alpha \varphi =_{df} \bigvee_{i \in Role} (F^i \varphi \wedge AS_\alpha(i))$

Proposition 3. *For any* $\alpha \in Agt$, $\models_{\mathbf{MR}} O_\alpha \varphi \rightarrow P_\alpha \varphi$ *and* $\models_{\mathbf{MR}} O_\alpha \neg \varphi \leftrightarrow F_\alpha \varphi$.

Proof. Let $\mathcal{M}$ be any model in **MR**. Suppose $\mathcal{M} \models O_\alpha\varphi$, then there is a role $r \in Role$ such that $\mathcal{M} \models O^r\varphi \wedge AS_\alpha(r)$. Thus, $R_r(w) \subseteq [\![\varphi]\!]$ and $R_r(w) \neq \emptyset$. It follows that $\mathcal{M} \models O^r\varphi \wedge AS_\alpha(r) \rightarrow \neg O^r\neg\varphi \wedge AS_\alpha(r)$. Since $P^r\varphi \leftrightarrow \neg O^r\neg\varphi$, we have $\mathcal{M} \models O^r\varphi \wedge AS_\alpha(r) \rightarrow P^r\varphi \wedge AS_\alpha(r)$. Hence, by definition of $P_\alpha\varphi$, we have $\mathcal{M} \models O_\alpha\varphi \rightarrow P_\alpha\varphi$.

By the definition of F^i, we have $\models_{\mathbf{MR}} O^i\neg\varphi \wedge AS_\alpha(i) \leftrightarrow F^i\varphi \wedge AS_\alpha(i)$. It follows easily that $\models_{\mathbf{MR}} O_\alpha\neg\varphi \leftrightarrow F_\alpha\varphi$. □

Proposition 4. *For any $\alpha \in Agt$, $O_\alpha\varphi \wedge O_\alpha\neg\varphi$, $O_\alpha\varphi \wedge P_\alpha\neg\varphi$ and $O_\alpha\varphi \wedge F_\alpha\varphi$ are satisfiable.*

Proof. Given a **MR**-model $\mathcal{M} = \langle W, R, (R_i)_{i\in Role}, C, V\rangle$ and a possible ideal world $w \in W$. Suppose that $\mathcal{M}, w \models O^j\varphi \wedge O^k\neg\varphi$ and $C(w,\alpha) = \{j,k\}$. Then, we have $\mathcal{M}, w \models AS_\alpha(j) \wedge O^j\varphi$ and $\mathcal{M}, w \models AS_\alpha(k) \wedge O^k\neg\varphi$. By Proposition 1, it follows that $\mathcal{M}, w \models O_\alpha\varphi \wedge O_\alpha\neg\varphi$. The proof of the other two cases is similar. □

Based on Proposition 4, we can conclude that any normative conflict is satisfiable.

4 Axiomatization

In this section, we present a Hilbert-style system **S** for the logic **DL-MR** and prove the corresponding soundness and completeness theorems.

Axiom Schemata:

(A1) All propositional tautologies

(A2) $O(\varphi \rightarrow \psi) \rightarrow (O\varphi \rightarrow O\psi)$

(A3) $O\varphi \rightarrow \neg O\neg\varphi$

(A4) $O^i(\varphi \rightarrow \psi) \rightarrow (O^i\varphi \rightarrow O^i\psi)$ (for any $i \in Role$)

(A5) $O\varphi \rightarrow O^i\varphi$ (for any $i \in Role$)

(A6) $AS_\alpha(i) \rightarrow \neg O^i\bot$ (for any $i \in Role$ and any $\alpha \in Agt$)

(A7) $O_\alpha\varphi \leftrightarrow \bigvee_{i\in Role} (O^i\varphi \wedge AS_\alpha(i))$ (for any $\alpha \in Agt$)

Rules of Inference:

(Modus Ponens) from φ and $\varphi \rightarrow \psi$ infer ψ

(Necessitation of O) from φ infer $O\varphi$

(Necessitation of O^i) from φ infer $O^i\varphi$

A formula φ is a theorem of **S** (notation: $\vdash_{\mathbf{S}} \varphi$), if there is a proof of φ in **S**. Let $Thm(\mathbf{S})$ be the set of all theorems of **S**.

Lemma 1. *The following formulas are theorems of **S**:*

(1) $O^i\varphi \wedge O^i\psi \to O^i(\varphi \wedge \psi)$
(2) $AS_\alpha(i) \to \neg O\bot$
(3) $O_\alpha\varphi \to \neg O\neg\varphi$
(4) $\neg O_\alpha\bot$

We say a formula φ is **S**-consistent, if $\nvdash_{\mathbf{S}} \neg\varphi$. A finite set of formulas Σ is **S**-consistent if the conjunction of all members of Σ is **S**-consistent, and a infinite set Σ of formulas is **S**-consistent if all of its finite subsets are **S**-consistent. Furthermore, a set of formulas Λ is a maximal **S**-consistent set if for any $\varphi \notin \Lambda$, $\Lambda \cup \{\varphi\}$ is not **S**-consistent. Via the standard argument of Lindenbaum's Lemma, every **S**-consistent set Λ can be extended to a maximal **S**-consistent set Λ^+.

Lemma 2. *Let Λ^+ be a maximal **S**-consistent set, then:*

(1) $\bot \notin \Lambda^+$,
(2) if $\varphi, \varphi \to \psi \in \Lambda^+$, *then* $\psi \in \Lambda^+$,
(3) for any formula φ, *exactly one of* φ *and* $\neg\varphi$ *is in* Λ^+,
(4) $\varphi \vee \psi \in \Lambda^+$ *iff* $\varphi \in \Lambda^+$ *or* $\psi \in \Lambda^+$,
(5) $Thm(\mathbf{S}) \subseteq \Lambda^+$.

Definition 5. *The canonical **MR**-model* $\mathcal{M}^c = \langle W^c, R^c, (R_i^c)_{i \in Role}, C^c, V^c \rangle$ *for **S** is defined as follows:*

- $W^c = \{w : w$ *is a maximal **S**-consistent set* $\}$;
- $R^c = \{(w, w') \in W^c \times W^c :$ *for any formula* φ, *if* $O\varphi \in w$, *then* $\varphi \in w'\}$;
- $R_i^c = \{(w, w') \in W^c \times W^c :$ *for any formula* φ, *if* $O^i\varphi \in w$, *then* $\varphi \in w'\}$, *for any* $i \in Role$;
- $C^c : W^c \times Agt$ *to* $\mathcal{P}(Role)$ *is a function s.t. for any pair* $(w, \alpha) \in W^c \times Agt$, $C^c(w, \alpha) = \{i \in Role : AS_\alpha(i) \in w\}$;
- $V^c(p) = \{w \in W^c : p \in w\}$, *for any atomic proposition* $p \in Atm$.

Lemma 3. *(Existence Lemma) For any element* $w \in W^c$, *if* $\neg O\neg\varphi \in w$, *then there is an element* $w' \in W^c$ *such that* $(w, w') \in R^c$ *and* $\varphi \in w'$. *Besides, for any element* $w \in W^c$ *and any* $i \in Role$, *if* $\neg O^i\neg\varphi \in w$, *then there is an element* $w' \in W^c$ *such that* $(w, w') \in R^i$ *and* $\varphi \in w'$.

Proof. The proof of the existance lemma is as in standard normal modal logic. We leave the details to the reader. □

Lemma 4. *The canonical **MR**-model* $\mathcal{M}^c$ *is a **MR**-model.*

Proof. We can check the serial property of R^c by A3: $O\varphi \to \neg O\neg\varphi$ and the Existence Lemma. The case for V^c is routine. Now we only to prove (1) R_i^c is a subset of R^c for any $i \in Role$ and (2) C^c is a role function.

The proof of (1): Suppose that $(w, w') \in R_i^c$. Then, by definition of R_i^c, we have $\{\varphi : O^i\varphi \in w\} \subseteq w'$. Since $O\varphi \to O^i\varphi$ is an axiom of **S**, it follows that $\{\psi : O\psi \in w\} \subseteq \{\varphi : O^i\varphi \in w\} \subseteq w'$. By definition of R^c, we have $(w, w') \in R^c$.

The proof of (2): Given a set w in W^c. Suppose that $AS_\alpha(i) \in w$. Since w is a maximal **S**-consistent set and $AS_\alpha(i) \to \neg O^i\bot$ is an axiom of **S**, we have $AS_\alpha(i) \to \neg O^i\bot \in w$. Then, $\neg O^i\bot \in w$. By the Existence Lemma, there is an element $w' \in W^c$ such that $(w, w') \in R_i^c$. Hence, $R_i^c(w) \neq \emptyset$. □

Lemma 5. *(Truth Lemma) Let $\mathcal{M}^c = \langle W^c, R^c, (R_i^c)_{i \in Role}, C^c, V^c \rangle$ be the canonical* ***MR****-model, and let $w \in W^c$. For any formula $\varphi \in \mathcal{L}$, $M^c, w \models \varphi$ iff $\varphi \in w$.*

Proof. The proof goes by induction on the structure of φ. The arguments of the atomic case and the boolean cases are standard. We only consider the other cases.

(Case $\varphi = AS_\alpha(i)$)

Suppose that $AS_\alpha(i) \in w$. By definition of C^c, we have $i \in C^c(w, \alpha)$. It follows that $\mathcal{M}, w \models AS_\alpha(i)$. On the other hand, $AS_\alpha(i) \notin w$. Then, by definition of C^c, we have $i \notin C^c(w, \alpha)$. So $\mathcal{M}, w \not\models AS_\alpha(i)$.

(Case $\varphi = O^i\psi$)

Suppose that $O^i\psi \in w$. Then, by definition of R_i, we have $\psi \in w'$ for any $w' \in R_i(w)$. By the inductive hypothesis, we have $\mathcal{M}, w' \models \psi$ for any $w' \in R_i(w)$. Hence, $\mathcal{M}, w \models O^i\psi$. Now for the other direction suppose that $O^i\psi \notin w$. First, we show that $\{\neg\psi\} \cup \{\tau : O^i\tau \in w\}$ is **S**-consistent. To arrive at a contradiction, assume that $\{\neg\psi\} \cup \{\tau : O^i\tau \in w\}$ is not **S**-consistent. Then there is a finite set of formulas $\{\tau_1, \ldots, \tau_n\} \subseteq \{\tau : O^i\tau \in w\}$ such that $\vdash_{\mathbf{S}} (\tau_1 \wedge \cdots \wedge \tau_n) \rightarrow \psi$. Thus, we have $\vdash_{\mathbf{S}} O^i(\tau_1 \wedge \cdots \wedge \tau_n) \rightarrow O^i\psi$ by the necessitation of O^i and A4. By Lemma 1(1), we would get $\vdash_{\mathbf{S}} (O^i\tau_1 \wedge \cdots \wedge O^i\tau_n) \rightarrow O^i(\tau_1 \wedge \cdots \wedge \tau_n)$. Hence, we have $\vdash_{\mathbf{S}} (O^i\tau_1 \wedge \cdots \wedge O^i\tau_n) \rightarrow O^i\psi$. Since $\{O^i\tau_1, \ldots O^i\tau_n\} \subseteq w$, it follows that $O^i\psi \in w$, a contradiction. Hence, $\{\neg\psi\} \cup \{\tau : O^i\tau \in w\}$ is **S**-consistent. So there is a maximal **S**-consistent set $\Sigma \in W^c$ such that $\{\neg\psi\} \cup \{\tau : O^i\tau \in w\} \subseteq \Sigma$. Thus, we have $\Sigma \in R_i(w)$. By the induction hypothesis, it follows that $\mathcal{M}, \Sigma \models \neg\psi$. Consequently, $\mathcal{M}, w \not\models O^i\psi$.

(Case $\varphi = O\psi$)

The proof is similar to the case for $\varphi = O^i\psi$.

(Case $\varphi = O_\alpha\psi$)

Suppose that $O_\alpha\psi \in w$. Since $O_\alpha\varphi \leftrightarrow \bigvee_{i \in Role} (O^i\varphi \wedge AS_\alpha(i))$ is an axiom of **S**, we have $\bigvee_{i \in Role} (O^i\varphi \wedge AS_\alpha(i)) \in w$. Then there is at least a $r \in Role$ such that $O^r\psi \wedge AS_\alpha(r) \in w$. By the above cases, $\mathcal{M}, w \models O^r\psi \wedge AS_\alpha(r)$. It follows that $\mathcal{M}, w \models O_\alpha\psi$. For the other direction suppose that $O_\alpha\psi \notin w$. Then, for any $j \in Role$, $O^j\psi \wedge AS_\alpha(j) \notin w$. By the above cases, $\mathcal{M}, w \not\models O^j\psi \wedge AS_\alpha(j)$ for any $j \in Role$. Hence, $\mathcal{M}, w \not\models O_\alpha\psi$. □

Theorem 1. *The logic* ***S*** *is sound and complete with respect to models in* ***MR***.

Proof. To prove soundness is a routine process. The soundness of **S** can be proved by proving that all axioms are valid and proving that all inference rules are preserve-validity. The completeness proof proceeds as follows:

Suppose that φ is not a theorem of **S**. Then $\neg\varphi$ is consistent with **S**. Hence, for the canonical **MR**-modal $\mathcal{M}^c$, there is a maximal **S**-consistent set $u \in W^c$ such that $\neg\varphi \in u$. It follows from the Truth Lemma that $\mathcal{M}, w \not\models \varphi$. Consequently, φ is not valid. □

Corollary 1. ***DL-MR*** $= Thm(\mathbf{S})$.

5 Decidability

In this section, our aim is to prove the decidability of the logic **DL-MR**. For this, we shall show that the logic **DL-MR** has the finite model property. In technical aspect, we use a method similar to filtration.

Let Γ denote a subformulas closed set of formulas. Note that if $O_\alpha\varphi \in \Sigma$, then Γ contains all subformulas of $\bigvee_{i\in Role} (O^i\varphi \wedge AS_\alpha(i))$.

Definition 6. *Given a* ***MR****-model* $\mathcal{M} = \langle W, R, (R_i)_{i\in Role}, C, V\rangle$ *and a subformulas closed set of formulas* Γ. *For all* $w, w' \in W$, *an equivalence* $\equiv_\Gamma$ *is defined as follows:*

$w \equiv_\Gamma w'$ *if and only if for all* $\varphi \in \Gamma$: $(\mathcal{M}, w \vDash \varphi$ *if and only if* $\mathcal{M}, w' \vDash \varphi)$.

Based on the above definition, we can introduce the following definition.

Definition 7. *Given a* ***MR****-model* $\mathcal{M} = \langle W, R, (R_i)_{i\in Role}, C, V\rangle$ *and a subformulas closed set of formulas* Γ. *A filtration model of* $\mathcal{M}$ *through* Γ *is a tuple* $\mathcal{M}^f = \langle W^f, R^f, (R_i^f)_{i\in Role}, C^f, V^f\rangle$ *where:*

- $W^f = \{[w] : w \in W\}$, $[w]$ *is the equivalence class of* w *with respect to the equivalence relation* $\equiv_\Gamma$;
- $R^f = \{([w], [u]) :$ *for some* $w' \in [w]$ *and* $u' \in [u], (w', u') \in R\}$;
- $R_i^f = \{([w], [u]) :$ *for some* $w' \in [w]$ *and* $u' \in [u], (w', u') \in R_i\}$;
- $C^f : W^f \times Agt \to \mathcal{P}(Role)$ *is a function such that for any pair* $([w], \alpha)$ *in* $W^f \times Agt$ *and any* $AS_\alpha(i) \in \Gamma$, $C([w], \alpha) = \{i \in Role : \mathcal{M}, w \vDash AS_\alpha(i)\}$;
- $V^f(p) = \{[w] : w \in V(p)\}$, *for any atomic proposition* $p \in \Gamma$.

Proposition 5. W^f *contains at most* $2^{|\Gamma|}$ *elements.*

Proposition 6. $\mathcal{M}^f$ *is a* ***MR****-model.*

Proof. We only need to prove (1) R_i^f is a subset of R^f for any $i \in Role$ and (2) C^f is a role function.

The proof of (1): Suppose that $([w], [u]) \in R_i^f$. By definition of R_i^f, there exists $w' \in [w]$ and $u' \in [u]$ such that $(w', u') \in R_i$. Since $R_i \subseteq R$, it follows that $(w', u') \in R$. Then, by definition of R^f, we have $([w], [u]) \in R^f$.

The proof of (2): Let $[w]$ be any element of W^f. Suppose $\mathcal{M}^f, [w] \vDash AS_\alpha(i)$. Then, by definition of C^f, we have $\mathcal{M}, w \vDash AS_\alpha(i)$. Since $AS_\alpha(i) \to \neg O^i\bot$ is a valid formula, we have $\mathcal{M}, w \vDash \neg O^i\bot$. Hence, there is a possible ideal world $u \in W$ such that $(w, u) \in R_i$. By definition of R_i^f, $([w], [u]) \in R_i^f$. So we have $R_i^f([w]) \neq \emptyset$. □

Theorem 2. *Given a* ***MR****-model* $\mathcal{M} = \langle W, R, (R_i)_{i\in Role}, C, V\rangle$ *and a subformulas closed set of formulas* Γ. *Let* $\mathcal{M}^f$ *be a filtration of* $\mathcal{M}$ *through* Γ. *For all* $\varphi \in \Gamma$ *and all* $w \in W$,

$$\mathcal{M}, w \vDash \varphi \text{ if and only if } \mathcal{M}^f, [w] \vDash \varphi.$$

Proof. The proof is by induction on the structure of φ. We leave the details to the reader. □

Theorem 3. *Given a formula $\varphi \in \mathcal{L}$, if φ is satisfiable in a **MR**-model, then φ is also satisfiable in a finite **MR**-model.*

Proof. Suppose that φ is satisfiable in a **MR**-model. By the soundness of **S**, we have $\nvdash_{\mathbf{S}} \neg\varphi$. This implies that φ is satisfiable in the canonical **MR**-model $\mathcal{M}^c$. Then we can construct a filtration model of $\mathcal{M}^c$ through the set of all subformulas of φ. By Theorem 2 and Proposition 5, we obtain the desired result. □

Corollary 2. ***DL-MR** is decidable.*

6 Collective Obligation

This section discusses the collective obligation based on the concept of role. A role can be viewed as a testing way to identify a selected group, such as a group of police officers, a group of teachers, and a group of doctors. However, all members of a group do not necessarily hold the same role. Especially in real life, it is rare that all members act as the same role in a social organization. Furthermore, some members may even be assigned multiple roles in the organization. Hence, we propose two distinct modes to identify collective obligations. In the first mode, we do not care about differential roles between members in a group. A differential role means that some members hold it in a particular situation, while others do not. In the second mode, we consider the collective obligations by the clear roles of members in a group. The second mode allows for the differential roles. In our terminology, the collective obligations in the first mode are called the simple collective obligations, and the collective obligations in the second mode are called the identified collective obligations.

6.1 Simple Collective Obligation

In considering the simple collective obligation, we treat a group G as a set of agents, i.e., $G \subseteq Agt$. For any group G, we introduce an operator O_G to stand for the concept of the simple collective obligations of G. Besides, we stipulate that $G \neq \emptyset$ for any group G.

Definition 8. *Give a **MR**-model $\mathcal{M} = \langle W, R, (R_i)_{i \in Role}, C, V\rangle$, a possible ideal world $w \in W$ and a group G. The satisfaction condition for the operator O_G is defined as follows:*

$$\mathcal{M}, w \models O_G\varphi \text{ iff there exists a } r \in Role \text{ s.t. } \mathcal{M}, w \models \bigwedge_{\alpha \in G} AS_\alpha(r) \text{ and } \mathcal{M}, w \models O^r\varphi$$

The underlying idea of the above definition is to consider whether there is a shared role between all members of group G. If all members can reasonably act as the same role i, then the role-related obligations placed on role i become the collective obligations of group G.

Based on the satisfaction condition for O_G, the formula $O_G\varphi$ can be defined by the following formula:

$$O_G\varphi =_{df} \bigvee_{i \in Role} (\bigwedge_{\alpha \in G} AS_\alpha(i) \wedge O^i\varphi).$$

Since the formula $O^i\varphi \wedge AS_\alpha(i) \to O_\alpha\varphi$ can be derived in **S**, we have the following formula:

$$O_G\varphi \to \bigwedge_{\alpha \in G} O_\alpha\varphi.$$

Next, we start to discuss some properties of the simple collective obligation. According to the definition of O_G, the operator O_G is tolerant of normative conflicts. Consider the below example.

Example 1. Given a **MR**-model $\mathcal{M} = \langle W, R, (R_i)_{i \in Role}, C, V \rangle$ and a possible ideal world $w \in W$. Assume that $\mathcal{M}, w \vDash O^j\varphi \wedge O^r\neg\varphi$. For a group $G = \{\alpha, \beta\}$, if $\mathcal{M}, w \vDash AS_\alpha(j) \wedge AS_\beta(j) \wedge AS_\alpha(r) \wedge AS_\beta(r)$, then $\mathcal{M}, w \vDash O_G\varphi \wedge O_G\neg\varphi$.

Based on this example, we can say that if all members of a group do not have a differential role, then the group may be in a notorious situation that does not have a definite deontic constraint, i.e., any formula and its negation are simultaneously obligatory for the group.

By the satisfaction condition for O_G, it is easy to check that the operator O_G satisfies the property of reduction. This property can be represented as follows:

$$\text{If } G_2 \subseteq G_1, \text{ then } \vDash_{\mathbf{MR}} O_{G_1}\varphi \to O_{G_2}\varphi.$$

Besides, we can find out that the operator O_G does not satisfy the property of merging. It means that if φ is obligatory for a group G_1 and ψ is obligatory for a group G_2, we cannot obtain that $\varphi \wedge \psi$ is obligatory for the combination $G_1 \cup G_2$ when $G_1 \nsubseteq G_2$ and $G_2 \nsubseteq G_1$. This fact can be represented as follows:

$$\text{If } G_1 \nsubseteq G_2,\ G_2 \nsubseteq G_1, \text{ and } \nvDash_{\mathbf{MR}} \varphi \leftrightarrow \psi,$$
$$\text{then } O_{G_1}\varphi \wedge O_{G_2}\psi \nvDash_{\mathbf{MR}} O_{G_1 \cup G_2}\varphi \wedge \psi.$$

Thus, we can obtain that the operator O_G also does not satisfy the property of expansion. This fact can be represented as follows:

$$\text{If } G_1 \subset G_2, \text{ then } O_{G_1}\varphi \nvDash_{\mathbf{MR}} O_{G_2}\varphi.$$

Since the operator O_G does not satisfy the property of merging and the property of expansion, we can say that increasing the number of members in a group possibly affects the original simple collective obligations of the group.

6.2 Identified Collective Obligation

Identifying what roles each member of a group is assigned is very useful for understanding the group. Hence, a group can be defined as a combination of different agents assigned roles. In considering the identified collective obligation, we treat a group G^* as a set of formulas in the form $AS_\alpha(i)$. For example, $\{AS_\alpha(r), AS_\beta(j)\}$ is a group in considering the identified collective obligation.

For any group G^*, we introduce an operator O_{G^*} to stand for the concept of the identified collective obligations of G^*. Besides, we stipulate that $G^* \neq \emptyset$ for any group G^*. We use $G^*_{Agt} = \{\alpha \in Agt : \exists i \in Role \text{ s.t. } AS_\alpha(i) \in G^*\}$ to denote the set of all agents in G^* and use $G^*_{Role} = \{i \in Role : \exists \alpha \in Agt \text{ s.t. } AS_\alpha(i) \in G^*\}$ to denote the set of all roles in G^*.

Definition 9. *Give a* ***MR****-model* $\mathcal{M} = \langle W, R, (R_i)_{i \in Role}, C, V\rangle$, *a possible ideal world* $w \in W$ *and a group* G^*. *The satisfaction condition for the operator* O_{G^*} *is defined as follows:*

$$\mathcal{M}, w \vDash O_G\varphi \quad \text{iff} \quad \mathcal{M}, w \vDash \bigwedge_{i \in G^*_{Role}} O^i\varphi \text{ and } \mathcal{M}, w \vDash \bigwedge G^*$$

The basic idea of the above definition is to consider whether there is a shared obligation for all roles in the group. If there exists a shared obligatory and all agents in the group can reasonably act as their roles, then the shared obligation become a collective obligation of the group.

Based on the satisfaction condition for O_{G^*}, the formula $O_{G^*}\varphi$ can be defined by the following formula:

$$O_{G^*}\varphi =_{df} \bigwedge_{i \in G^*_{Role}} O^i\varphi \wedge \bigwedge G^*.$$

Since the formula $O^i\varphi \wedge AS_\alpha(i) \to O_\alpha\varphi$ can be derived in **S**, we have the following formula:

$$O_{G^*}\varphi \to \bigwedge_{\alpha \in G^*_{Agt}} O_\alpha\varphi.$$

Now, we start to discuss some properties of the identified collective obligation. By the definition of $O_{G^*}\varphi$, we can infer that

$$\neg O_{G^*}\neg\varphi \leftrightarrow \bigvee_{i \in G^*_{Role}} \neg O^i\neg\varphi \vee (\neg \bigwedge G^*).$$

Then, we obtain an ideal result:

$$\vDash_{\mathbf{MR}} O_{G^*}\varphi \to \neg O_{G^*}\neg\varphi.$$

Hence, the operator O_{G^*} is not tolerant of normative conflicts. Based on this property, we can say that when all members of a group at least hold a clear role, the group cannot tolerate normative conflicts.

Like the operator O_G, the operator O_{G^*} satisfies the property of reduction. This can be represented as follows:

$$\text{If } G_2^* \subseteq G_1^*, \text{ then } \models_{\mathbf{MR}} O_{G_1^*}\varphi \to O_{G_2^*}\varphi.$$

Moreover, the operator O_{G^*} also does not satisfy the property of merging. This fact can be represented as follows:

$$\text{If } G_1^* \nsubseteq G_2^*,\ G_2^* \nsubseteq G_1^*, \text{ and } \nvDash_{\mathbf{MR}} \varphi \leftrightarrow \psi,$$
$$\text{then } O_{G_1^*}\varphi \wedge O_{G_2^*}\psi \nvDash_{\mathbf{MR}} O_{G_1^* \cup G_2^*}\varphi \wedge \psi.$$

Thus, we can obtain that the operator O_{G^*} does not satisfy the property of expansion. This fact can represented as follows:

$$\text{If } G_1^* \subset G_2^*, \text{ then } O_{G_1^*}\varphi \nvDash_{\mathbf{MR}} O_{G_2^*}\varphi$$

Since the operator O_G^* does not satisfy the property of merging and the property of expansion, we can say that increasing the number of members in a group possibly affects the original identified collective obligations of the group. Besides, by the definition of $O_{G^*}\varphi$, we can know that if the roles in a group are changed, then the original identified collective obligations of the group might be affected.

7 Conclusion and Future Works

We think that the role-related deontic syllogism is a kind of practical reasoning and is purposed to ensure the transformation from the role-related obligation to the individual obligation. The premise of applying a role-related deontic syllogism in a practical scenario is that the roles held by the agent are reasonable in practice. Based on the role-related deontic syllogism, the deontic logic of multiple roles(DL-MR) is proposed to characterize the normative conflict arising from the role-related obligations. From the syntactical perspective, it is noteworthy that the form $AS_\alpha(i)$ are dissimilar from those general forms of formulas in modal logic. The two modes of collective obligations discussed in this paper are all considered from the perspective of role-related obligations, which cannot reflect all characteristics of collective obligations. We believe that the collective obligations are not only determined by role-related obligations but also closely associated with collective capabilities and tasks in various circumstances. Based on the present work, we may consider the following further directions:

Public announcement of roles: In daily life, it is a common practice to obtain deontic support for actions by announcing reasonable roles in practical scenarios. For example, when a police officer asks a suspect to cooperate in an investigation, the police officer will first show his/her identity document to clarify that he/she is a police officer. The public announcement of roles can determine what role-related obligations ought to be fulfilled between specific agents. In addition, by adding the knowledge factor, we can analyze whether the agent knows the announced role, so as to evaluate the effectiveness of the announcement.

Preference for roles: When two role-related obligations lead to a normative conflict, ignoring one of the two role-related obligations can eliminate the conflict. An appropriate way of ignoring roles is to introduce a preference relation between roles. Depending on the preference relation, the agent can only consider the most preferred role and ignore the other roles.

References

1. Calardo, E., Rotolo, A.: Variants of multi-relational semantics for propositional non-normal modal logics. Journal of Applied Non-Classical Logics **24**(4), 293–320 (2014)
2. Chellas, B.F.: Modal logic: an introduction. Cambridge university press (1980)
3. Donagan, A.: Consistency in rationalist moral systems. The Journal of Philosophy **81**(6), 291–309 (1984)
4. Goble, L.: Preference semantics for deontic logic part i—simple models. Logique et Analyse **46**(183-184), 383–418 (2003)
5. Goble, L.: Preference semantics for deontic logic part ii–multiplex models. Logique et Analyse **47**(185-188), 335–363 (2004)
6. Pacheco, O., Santos, F.: Delegation in a role-based organization. In: International Workshop on Deontic Logic in Computer Science. pp. 209–227. Springer (2004)
7. Pacuit, E.: Neighborhood semantics for modal logic. Springer (2017)
8. Sartre, J.P.: Existentialism Is a Humanism. Yale University Press (2007)
9. Van der Torre, L., Hulstijn, J., Dastani, M., Broersen, J.: Specifying multiagent organizations. In: International Workshop on Deontic Logic in Computer Science. pp. 243–257. Springer (2004)

EEG Evidence for Game-Theoretic Model to Ambiguous Pronoun Resolution*

Mengyuan Zhao[1] [0000–0002–8917–0780], Zhong Yin[2,3] [0000–0003–2013–4009], and Ziming Lu[3] [0000––0002–6215–3745]

[1] School of Marxism, University of Shanghai for Science and Technology, Shanghai 200093, PR China
mengyuan.zhao@usst.edu.cn
[2] Engineering Research Center of Optical Instrument and System, Ministry of Education, Shanghai Key Lab of Modern Optical System, University of Shanghai for Science and Technology, Shanghai 200093, PR China
yinzhong@usst.edu.cn
[3] Department of Linguistics and Translation, School of International Studies, Zhejiang University, Hangzhou, 310058, PR China
zimінglu@zju.edu.cn

Abstract. In this paper we develop a game-theoretic model of ambiguous pronoun resolution, namely, the pronoun reference is not clearly determined in the context. We propose that iterated best response (IBR) reasoning offers a reasonable solution to ambiguous pronoun processing. Using electroencephalogram (EEG) (14 channels) to investigate Chinese processing, we provide evidence that the processes of resolving ambiguous and unambiguous pronouns are significantly different at both neural and behavioural level. The differences mainly manifest in longer reaction time and signals collected from the channels O1 (left occipital cortex) and P8 (right inferior parietal cortex), which are activated during probabilistic expected utility generation. These findings are consistent with general assumptions of our model that ambiguous pronoun resolution involves a mechanism of decision-making.

Keywords: Pronouns · Game theory · EEG.

1 Introduction

Personal pronouns such as *he* and *she* refer to an earlier mentioned person in the context. Pronoun resolution is a fundamental process in daily language processing, and many linguistic studies have investigated how people assign a referent to a pronoun according to grammatical rules (see [1–4]). These researches have shown that some types of linguistic cues can be used to pronoun resolution including: gender, verb-bias, focus

* Supported by the National Social Science Fund under Grant No. 18CZX014, the Shanghai Philosophy and Social Sciences Fund under Grant No. 2017EZX008, the National Natural Science Foundation of China under Grant No. 61703277, Shanghai Sailing Program under Grant No. 17YF1427000. This chapter is in its final form and it is not submitted to publication anywhere else.

B. Liao and Y. N. Wáng (eds.), *Context, Conflict and Reasoning*,
Logic in Asia: Studia Logica Library,
https://doi.org/10.1007/978-981-15-7134-3_4

and so on. However, pronouns carry little content by nature, and thus are referentially ambiguous in certain cases. Compare, for example, the following sentences:

(1) The wife stopped the husband. She cried.
(2) The owner blamed the waiter. He was angry.

In (1), the gender information of the nouns help to identify pronoun denotation, and therefore the pronoun *she* must refer to *the wife*. In this case, the pronoun's referent is determined by some linguistic cues (here, gender), and we call it *the unambiguous pronoun*. While in (2), gender is not informative to determine the pronoun's referent, we call it *the ambiguous pronoun*. The pronoun *he* in (2) may either refer to *the owner* or *the waiter*, and its referent is undetermined. Another linguistic cue that may help pronoun resolution is the verb-bias, that is, the semantic meaning of the verb may lead to a bias for pronoun interpretation. In a seminal work of psycholinguistics on pronouns, Garvey & Caramazza [1] have reported strong bias in pronoun resolution for specific verbs. They found a strong bias in interpreting *she* as a reference of the object *the daughter* for the verb *scold* (e.g. *The mother scolded her daughter becasuse she* ...), while *she* referring to the subject *the mother* for the verb *confess* (e.g. *The mother confessed to her daughter becasuse she* ...). This suggests that people do not simply decide the referent of an ambiguous pronoun by proximity.

Unambiguous expressions show prima facie advantages comparing to their ambiguous counterparts. Actually, the pronoun *he* as shown in sentence (2) is more brief than the definite descriptions *the owner* or *the waiter*. Brevity is commonly argued as a rationality principle in communication. Grice's [5] Maxims of Conversation have shown this tension between brevity and ambiguity: Maxim of Quantity and Maxim of Manner. Therefore, it is worthy of discussion about the rationale of the use of ambiguous expressions in daily communication. The present work assumes that ambiguous pronoun processing involves rational decision-making, which can be modelled in game theory. The tradition of applying game models to pragmatics can be traced back to the seminal work of David Lewis [6]. Lewis introduced signalling games, which characterize communication as a speaker's attempt to influence a hearer's action by sending a certain signal. On the basis of Lewisian tradition, Parikh [7, 8] developed a more comprehensive framework named *games of partial information*. Parikhian model has been applied to analyze reference resolution (see [9, 10]). Clark and Parikh [10] adopted the conception of *Pareto Nash Equilibrium* to solve the game of ambiguous reference. Another branch of game-theoretic pragmatics couches iterative dynamics as an analysis of rational language use (see for example [11–13]). An influential model of iterated response reasoning is the so called *pragmatic back-and-forth reasoning* developed by Franke and Jäger [13]. Recently, the dynamic reasoning model has been applied to an analysis of ambiguous expressions [14].

In this paper, we develop a game-theoretic model of ambiguous pronoun resolution. The model includes two parts: a signalling game as an illustration of the situations where people process ambiguous pronouns, and a reasoning account as a solution of the game. We point out that the solution conception of Parikhian model requiring the agents being rational enough to select the most profitable strategies is too constrictive. To solve the game of ambiguous pronoun, we introduce the iterated best response (IBR) reasoning. We argue that IBR reasoning by assuming a step-by-step interactive reasoning procedure allows analyzing actions under bounded rationality.

It is arguable whether the cognitive processing of pronoun resolution only involves the core language network or it also recruits the network of strategic decision-making as suggested in our model . To test the assumptions of our model, we use electroencephalogram (EEG) (14 channels) to study ambiguous pronoun processing in Chinese. Though recent years some work has been done on neural measures of reference resolution (see[15–19]), neural correlates to ambiguous pronoun resolution remain greatly uncharacterized. In our experiments, participants observed a sentence containing two nouns followed by the other sentence containing a personal pronoun (for example, sentence (1)). Behavioural data including reaction time (the time consumed by a participant to identify a pronoun referent) and neural data including EEG signals of 4 frequency bands and 14 channels during the whole resolution procedures are recorded. These data suggest a significant difference between processes of ambiguous and unambiguous pronoun resolution. The neural data of channel P8 and O1 demonstrate more activated recruitment of right inferior parietal cortex and left occipital cortex during ambiguous pronoun resolution. According to a review of neuroimaging studies (see [20, 21, 18]), these areas respectively implicate expected utility calculation under probabilistic situations and extra effort paid for entire sentence reading. These findings are consistent with the assumptions of our model that ambiguous pronoun resolution involves a mechanism of decision-making. Our work also extends previous research on English processing to Chinese processing, which is structurally different from English. The results indicate that game-theoretic model can be applied across languages, and encourage further generalization of the model in future research.

2 A Game-Theretic Model to Ambiguous Pronoun Resolution

In this section, we will first construct a game-theoretic model, and then apply it to the case of ambiguous pronoun resolution. The model follows the main assumptions from the tradition of Gricean pragmatics (see [5]): communication is considered to be a cooperative and rational activity. The model presented in this paper will consider about a basic case that involves just one ambiguous message. The current model is aiming at offering a brief guideline to analyze ambiguous pronoun resolution and is competent to explain the rationale of the pragmatics in cases such as sentence (2). To analyze more complicated sentences with more than one ambiguous pronouns would require an extended model which can be derived from the basic model in principle. However, the development of an extended model has gone beyond the present work.

2.1 The Model

Context Modelling In the model, we assume that there is a speaker S, who has the relevant information about the world where she is in, and a hearer H, who has to judge about the world by reasoning on the message that the speaker transmits. Assuming there are two possible worlds: w_1 and w_2, we now model the speaker's knowledge about the world as types: t_1 indicates that S knows that she is in w_1, t_2 indicates that S knows that she is in w_2. We introduce Nature, say N, as an impersonal player of the game. N chooses a move to either type, say $t \in T$, with a probability, say p. Let $Pr(t) \in \Delta(T)$ be prior distribution over types, where $\Delta(T)$ refers to a probability distribution over types $t_1, t_2, \ldots, t_n$. We assume that $p = Pr(t)$ implicates H's prior belief in t before receiving any message. S will send a message, say $m \in M$, to H to inform her about the world, and H will interpret the received message into a type of S. We assume that the players will play according to the sematic meaning of messages. This constraint can be shown by introducing a lexicon B that maps type-message pair to the Boolean truth-value of the message for the speaker's type. A minimal lexicon fragment that involves choices between ambiguous and unambiguous messages is one with two types and three messages, where $B(t_1, m_1) = B(t_1, \bar{m}) = 1$, $B(t_2, m_2) = B(t_2, \bar{m}) = 1$. Put into words, for type t_1, S may either send an unambiguous message containing a definite description, say m_1, or send an ambiguous message containing a pronoun, say $\bar{m}$. Similarly, for type t_2, S may send an unambiguous message m_2 or an ambiguous message $\bar{m}$. Accordingly, we assume that the speaker will play a semantically consistent strategy, say σ, which is defined as follows:

Definition 1. *A semantically consistent strategy of the speaker σ is a function that maps each speaker type $t \in T$ to a probability of each message $m \in M$ to be sent in t, given that m is semantically true in t: $\sigma \in \Delta(M)^T, B(t,m) = 1$.*

Thereafter, H will interpret m_1 into t_1, m_2 into t_2. And when H receives $\bar{m}$, she may possibly interpret it into either m_1 or m_2. Accordingly, we assume that the hearer will also play a semantically consistent strategy, say ρ, which is defined as follows:

Definition 2. *A semantically consistent strategy of the hearer ρ is a function that maps each message m to a probability of each interpretation t to be chosen, given that m is semantically true in t: $\rho \in \Delta(T)^M, B(t,m) = 1$.*

We further assume that a successful communication using a pronoun will provide players with an extra gain, say ε, since the communication is complete with less words. We assume that S and H are purely cooperative, in the sense that S and H share the same utility functions. It means that both S and H would prefer that H's interpretation of m, say t_j, is identical to S's type, say t_i. We define the utility functions of players as follows.

Definition 3. *Let $U_N(t_i, m, t_j)$ be payoff of $N \in \{S, H\}$ given t_i, m and t_j, where $i, j = 1, 2$.*

$$U_N(t_i, m, t_j) = \begin{cases} 1, & \text{if } i = j \text{ and } m \in \{m_1, m_2\} \\ 1+\varepsilon, & \text{if } i = j \text{ and } m = \overline{m} \\ 0, & \text{if } i \neq j \end{cases}.$$

Definition 3 suggests: Both players will gain a plain payoff, say 1, using unambiguous messages; both will gain an extra payoff, say 1+ε, if H correctly understands the ambiguous pronoun message; and both will earn 0, if H misinterprets the pronoun. We denote by $p \in (0,1)$ H's prior belief that S is of type t_1, i.e. $Pr(t_1) = p$. The game tree in Fig. 1 illustrates the signalling game of our model.

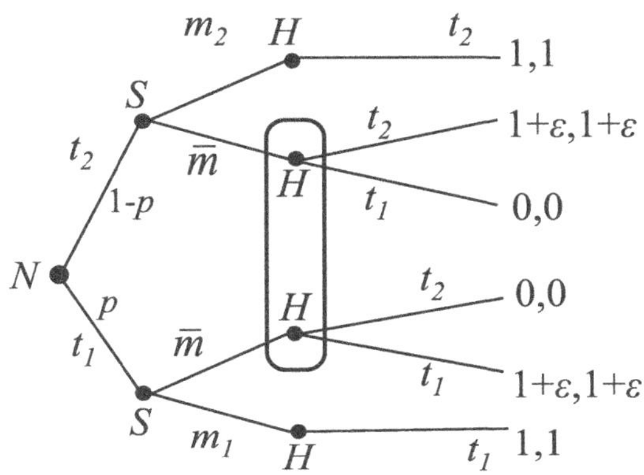

Fig. 1. A game tree for the model.

Solution Modelling As a solution of the game, we introduce the IBR reasoning framework. The IBR reasoning includes two reasoning sequences, namely, the S_0-sequence and the H_0-sequence. In the S_0-sequence, the first step of reasoning starts from a level-0 speaker, say S_0. We assume that S_0 is a naïve speaker, in the sense that she will choose a random message that is semantically consistent with her type. The level-1 hearer, say H_1 will play rationally based on her belief in S_0, in the sense that H_1 will choose the strategy that offers her the best expected utility. Similarly, S_2 will play rationally based on her belief in H_1, and the reasoning continues in this way. In the H_0-sequence, the first step will be taken by a level-0 hearer, say H_0. H_0 will choose the strategy that offers her the best expected utility based on her posterior belief in the speaker's strategy. In other words, we assume that H_0, unlike S_0, is rational. The level-1 speaker, say S_1 will play rationally according to her belief in H_0, and so on. As a generalization, a level-$(k+1)$ player will play rationally according to her belief in the strategies that a level-k player will choose. Now we illustrate by induction the IBR reasoning scaffolding.

Naïve Levels The S_0-sequence starts from the level-0 speaker S_0, who will randomly play a semantically consistent strategy. According to Definition 1, S_0 may choose either m_1 or $\bar{m}$ in t_1, and she may choose either m_2 or $\bar{m}$ in t_2. We perspicuously list all possible choices of S_0 for both types as follows.

$$S_0 = \left\{ \begin{array}{l} t_1 \mapsto m_1, \bar{m} \\ t_2 \mapsto m_2, \bar{m} \end{array} \right\}.$$

The H_0-sequence starts from the level-0 hearer H_0, who will play rationally according to her posterior belief. A posterior belief of H_0, say μ_0, results from the hearer's prior belief updated by the semantic meaning of the messages: $\mu_0(t|m) = Pr(t) \times B(t,m)$. For unambiguous messages m_1 and m_2, H_0 will choose semantic interpretations t_1 and t_2, respectively. For the ambiguous message $\bar{m}$, the posterior beliefs in two types are calculated as follows: $\mu_0(t_1|\bar{m}) = Pr(t_1) \times B(t_1,\bar{m}) = p$, $\mu_0(t_2|\bar{m}) = Pr(t_2) \times B(t_2,\bar{m}) = 1-p$. H_0 will choose any t with a higher posterior belief as an interpretation of $\bar{m}$. This means H_0's interpretation of the ambiguous message is dependent on the value of p, which represents the hearer's prior belief in speaker's types.

$$H_0 = \left\{ \begin{array}{l} \left\{ \begin{array}{l} m_1 \mapsto t_1, \\ m_2 \mapsto t_2, \quad \text{if } p > \frac{1}{2} \\ \bar{m} \mapsto t_1, \end{array} \right. \\ \left\{ \begin{array}{l} m_1 \mapsto t_1, \\ m_2 \mapsto t_2, \quad \text{if } p < \frac{1}{2} \\ \bar{m} \mapsto t_2, \end{array} \right. \end{array} \right\}.$$

Sophisticated Levels We assume that in both S_0- and H_0-sequences, level-$(k+1)$ player will play as the best response to her belief in the other's strategies ($k > 0$). For simplicity, we assume that the players believe that their opponents are of the level that is exactly one level lower than her own. For the level-$(k+1)$ speaker, her belief is given as the semantically consistent strategies (see Definition 2) of the level-k hearer, say ρ_k. We further assume that the speaker will give a best response to her belief, say $Br(\rho)$. A best response means a rational play, namely, the speaker will choose a pure strategy, say $s(t)$, that will maximize her expected utility. The speaker's expected utility, say $EU_S(t,m,\rho)$, can be calculated as follows.

$$EU_S(t,m,\rho) = \sum_{t_i \in T} \rho(t_i|m) \times U_S(t,m,t_i) \tag{1}$$

As shown in Equation (1), the expected utility of the speaker is a sum of the utilities (see Definition 3) of all possible outcomes weighted by their probabilities dependent on the speaker's belief. Accordingly, the speaker's strategy as a best response of her belief is as follows.

$$s(t) = BR(\rho) \in \arg\max_{m \in M} EU_S(t,m,\rho) \tag{2}$$

As for the level-$(k+1)$ hearer, her belief is given as the posterior belief, say μ_{k+1}, which is derived from the hearer's prior belief and the semantically consistent strategies (see Definition 1) of the level-k speaker, say σ_k, by Bayesian conditionalization.

$$\mu_{k+1}(t_j|m_i) = \frac{Pr(t_j) \times \sigma_k(m_i|t_j)}{\sum_{t' \in T} Pr(t') \times \sigma_k(m_i|t')} \tag{3}$$

In Equation (3), the Bayesian conditionalization represents the belief dynamics: the likelihood for each type t is computed after a certain message m is received in t given prior probability of t and the expected probability of m to be sent in t. We assume that the hearer will show a best response to her belief, say $Br(\mu)$. A best response of a hearer is a pure strategy, say $h(t)$, that will maximize her expected utility. The hearer's expected utility, say $EU_H(t,m,\mu)$, can be calculated as follows.

$$EU_H(t,m,\mu) = \sum_{t_i \in T} \mu_1(t_i|m) \times U_H(t_i,m,t) \tag{4}$$

In Equation (4), the hearer's expected utility is a sum of the utilities (see Definition 3) of all possible outcomes weighted by their probabilities dependent on the hearer's belief. Accordingly, the hearer's strategy as a best response of her belief is as follows.

$$h(m) = BR(\mu) \in \arg\max_{t \in T} EU_H(t,m,\mu) \tag{5}$$

Now we start reasoning in the S_0-sequence. After S_0 sends a certain m to H_1, the latter will respond based on her posterior belief in S_0, say $\mu_1(t|m)$. From Equations (3) and (4), we calculate H_1's expected utilities of different choices while receiving $\bar{m}$: $EU_{H_1}(t_1,\bar{m},\mu_1) = p \times (1+\varepsilon)$, $EU_{H_1}(t_2,\bar{m},\mu_1) = (1-p) \times (1+\varepsilon)$. From equation (5), H_1 will interpret $\bar{m}$ into t_1 if and only if $EU_{H_1}(t_1,\bar{m},\mu_1) > EU_{H_1}(t_2,\bar{m},\mu_1)$, requiring $p > \frac{1}{2}$. Thus, the strategy of H_1 can be perspicuously illustrated as follows.

$$H_1 = \left\{ \begin{array}{l} \left\{ \begin{array}{l} m_1 \mapsto t_1, \\ m_2 \mapsto t_2, \ \text{if } p > \frac{1}{2} \\ \bar{m} \mapsto t_1, \end{array} \right. \\ \left\{ \begin{array}{l} m_1 \mapsto t_1, \\ m_2 \mapsto t_2, \ \text{if } p < \frac{1}{2} \\ \bar{m} \mapsto t_2, \end{array} \right. \end{array} \right\}.$$

Accordingly, $h_1(m)$ can be calculated from Equation (5): $h_1(t_1|m_1) = h_1(t_2|m_2) = h_1(t_1|\bar{m}) = 1, \text{if } p > \frac{1}{2}$; $h_1(t_1|m_1) = h_1(t_2|m_2) = h_1(t_2|\bar{m}) = 1, \text{if } p < \frac{1}{2}$. S_2 will respond upon her belief, say ρ_1, which is equal to h_1. From equations (4) and (5), we illustrate S_2's strategies as follows.

$$S_2 = \left\{ \begin{array}{l} \left\{ \begin{array}{l} t_1 \mapsto \bar{m}, \\ t_2 \mapsto m_2, \end{array} \right. \text{if } p > \frac{1}{2} \\ \left\{ \begin{array}{l} t_1 \mapsto m_1, \\ t_2 \mapsto \bar{m}, \end{array} \right. \text{if } p < \frac{1}{2} \end{array} \right\}.$$

Accordingly, $s_2(t)$ can be computed from Equation (2): $h_1(\bar{m}|t_1) = h_1(m_2|t_2) = 1, \text{if } p > \frac{1}{2}$; $h_1(m_1|t_1) = h_1(\bar{m}|t_2) = 1, \text{if } p < \frac{1}{2}$. H_3 will give the best response upon her posterior belief in S_2, and her strategy can be figured out following a similar procedure of what happens in the case of H_1. We perspicuously illustrate H_3's strategy as follows.

$$H_3 = \left\{ \begin{array}{l} \left\{ \begin{array}{l} m_1 \mapsto t_1, \\ m_2 \mapsto t_2, \text{ if } p > \frac{1}{2} \\ \bar{m} \mapsto t_1, \end{array} \right. \\ \left\{ \begin{array}{l} m_1 \mapsto t_1, \\ m_2 \mapsto t_2, \text{ if } p < \frac{1}{2} \\ \bar{m} \mapsto t_2, \end{array} \right. \end{array} \right\}.$$

It is clear that H_3 shows the same strategic pattern as H_1. Given the reasoning principles of best response operation, S_4 should play the same as S_2, H_5 should play the same as H_3 and so on. This means the strategies of the players start to repeat themselves from the level-3 hearer after two rounds of best response reasoning in the S_0-sequence. And it is also easy to reach similar results in the H_0-sequence, of which we would like to skip the details for the sake of simplicity.

Predictions In our model, both the sets T and M are finite, therefore, there are finitely many pure strategies. This means the IBR reasoning sequences will definitely repeat themselves at a certain level. We define an idealized prediction of IBR reasoning as follows.

Definition 4. *The idealized predictions of IBR reasoning are infinitely repeated strategies S^* and H^*:*

$$\begin{aligned} S^* &= \{s \in S | \exists i \forall j > i : s \in S_j\} \\ H^* &= \{h \in H | \exists i \forall j > i : h \in H_j\} \end{aligned}$$

From the steps that we have shown in the S_0-sequence, the strategy repetition begins after two rounds of reasoning. And it is also easy to prove that a reasoning of H_0-sequence will lead to similar results. Accordingly, we provide a prediction of our model in Proposition 1:

Proposition 1.

$$S^* = \left\{ \begin{array}{l} \left\{ \begin{array}{l} t_1 \mapsto \bar{m}, \\ t_2 \mapsto m_2, \end{array} \right. \textit{if } p > \frac{1}{2} \\ \left\{ \begin{array}{l} t_1 \mapsto m_1, \\ t_2 \mapsto \bar{m}, \end{array} \right. \textit{if } p < \frac{1}{2} \end{array} \right\}. H^* = \left\{ \begin{array}{l} \left\{ \begin{array}{l} m_1 \mapsto t_1, \\ m_2 \mapsto t_2, \textit{ if } p > \frac{1}{2} \\ \bar{m} \mapsto t_1, \end{array} \right. \\ \left\{ \begin{array}{l} m_1 \mapsto t_1, \\ m_2 \mapsto t_2, \textit{ if } p < \frac{1}{2} \\ \bar{m} \mapsto t_2, \end{array} \right. \end{array} \right\}.$$

.

Proposition 1 suggests: Whether an ambiguous pronoun is used instead of an unambiguous noun is dependent on the hearer's prior belief in the frequency of the world where the speaker is in . When the hearer believes that it is more likely for the speaker be in the world w_1, where S is of type t_1, the speaker will send a pronoun message for t_1 and a noun message for t_2, and the hearer will successfully translate the pronoun message into t_1. And the same reasoning will also stand in the case where the hearer's prior belief is biased towards t_2.

2.2 Comparison with other Models

Gricean Approches The game-theoretic model of ambiguous pronoun resolution presented here is highly in the spirit of Gricean pragmatics (see [5]). Grice accounted for pragmatical reasoning as a rational behaviour of agents. Pronoun resolution has been explained by application of the Gricean or neo-Gricean approaches (see for example [22, 23]). Our game model follows Grice's idea by modelling inferences of ambiguous pronoun resolution as rational interaction between the speaker and the hearer.

The differences between our model and the Gricean approaches are mainly in two aspects. Firstly, in the conceptual aspect, the game model presented in this paper does not rely on a formulation of Grice's Maxim of Conversation. The game model is constructed based on a simple assumption of cooperation in the sense that both of the players share a common interest. And this cooperation is formalized in terms of the utility functions (see Definition 3). Furthermore, our model also leaves open the possibility of explaining non-cooperative situations by a revision of utility functions. Secondly, in the epistemic aspect, the model presented here uses an iterated best response reasoning to show epistemic concerns of which the Gricean approaches are lack. The IBR reasoning have three features: semantic meaning focus, step-by-step interactive pattern and tolerance of bounded rationality. These features correspond to actual epistemic situations. IBR reasoning starts from level-0 players, who select according to the semantic meaning of the messages. The semantic meaning acts as a psychological focus of the agents during the reasoning, that is, the agents are psychologically attracted by the semantic meaning, from which they start the pragmatic reasoning. The IBR model also simulates a step-by-step interactive reasoning. This framework allows agents to update their belief in each other's rational strategies, and to upgrade their reasoning level. In addition, IBR reasoning is tolerant to limited rationality, which shows a more real situation. The model offers freedom to stop at any level of sophistication to check the result of reasoning from either bounded or ideal rationality.

Games of Partial Information A game of partial information involves ambiguous information states in the game tree and is to be solved by adoption of Pareto-Nash Equilibrium (see [7, 8]). It has been applied to the analysis of pronoun resolution (see [9, 10]). Games of partial information share the same tradition of game-theoretic pragmatics with the model presented in this paper.

The difference between the model presented here and the games of partial information is mainly in the aspect of solution concepts. The games of partial information adopt the solution concept of Pareto-Nash Equilibrium. A Pareto-dominant Nash Equilibrium is the best-paid strategy profile among those which offer both players the best payoff given the strategy of her opponent. In other words, the Pareto-Nash Equilibrium is the most profitable equilibrium of the game. To follow this, it is required that the agents compute all equilibria and make the comparison as well. This requirement not only is too much for the rationality of the agents, but also presumes an outsider's view of the game to complete the calculation. In comparison, our model adopts the solution of IBR reasoning predictions. The IBR reasoning illustrates a step-by-step interaction. It allows the agents to respond from different levels of sophistication with more tolerance to the rationality of the players. It also shows as a simulation of the real procedure of the agents updating their belief and responding to it . Therefore, the IBR reasoning is more like an insider's view of the game.

Pragmatic Back-and-Forth Reasoning The back-and-forth reasoning combines the idea of signalling games as the context formulation and iterated response reasoning as the solution schemes (see [12, 13]). It has been applied to analyze the resolution of ambiguous reference (see [14]). The model presented here bear a close resemblance to the back-and-forth reasoning.

The difference between our model and the back-and-forth reasoning is mainly in two points. Firstly, as for the context modelling, the back-and-forth reasoning uses signalling games to describe the context of a sender sending messages to inform a receiver about the state t. Instead, in our model, we use t to represent the type of the speaker, who will send a message to inform the hearer about her knowledge of the world. Comparing to the settings of the back-and-forth reasoning, our model is capable of representing the speaker's expertise, and thus leaves open the possibility that the speaker has only partial information of the world. Secondly, as for the solution modelling, the back-and-forth reasoning includes at least three types of iterated response reasoning schemas: iterated best response, iterated cautious response and iterated quantal response. Our model adopts a solution concept that is most close to the iterated best response in Franke's work (see [12]). In the vanilla IBR model, Franke assumed that the receiver would show unbiased prior beliefs in all states. In comparison, our work intruduces a parameter p to represent the hearer's prior belief in different speaker types. This parameter is key to our model in the sense that it determines the final solution to the game. Furthermore, the parameter of prior belief also plays an important role both in our pretest work and in the experiments (see Section 3 for details).

2.3 The Application

Game-theoretic models have been applied to various pragmatic phenomenons (see [24–26] for a selective survey). However, most researches are based on an analysis of English sentences. We explore our game-theoretic model to an analysis of Chinese, which is structurally different from English. We first construct 200 pairs of Chinese sentences dividing into different groups, then we investigate the value of prior belief in the referent of ambiguous pronouns based on a 30-participant survey.

We construct 200 pairs of sentences through the following steps: We first identify 80 nouns, 40 of which are gender-biased (for example, *qizi* 'wife') and the other 40 gender-neutral (for example, *laoshi* 'teacher'); we then generate meaningful 200 sentences by pairing nouns with a transitive verb (for example, *piping* 'blame'); we finally generate 200 sentences including a pronoun and an intransitive verb (for example, *xiao* 'smile'). We divide the 200 pairs of sentences into different classes according to a group of characteristics. A main classification is to distinguish between ambiguous and unambiguous pronoun resolution. For example, compare the following pairs of sentences:

(3) Qizi lanzhu zhangfu. Ta ku-le.
wife stop husband. She cry ASP.
'The wife stopped the husband. She cried.'

(4) Dianzhang piping fuwuyuan. Ta shengqi-le.
owner blame waiter. He angry ASP.
'The owner blamed the waiter. He was angry.'

The pronoun in (3) unambiguously refers to *the wife*, while the pronoun in (4) may either refer to *the owner* or *the waiter*. Our model can be applied to analyze ambiguous pronoun resolution as shown in example sentences (4). Since the hearer's prior belief is key to solve the game as shown in Proposition 1, we conduct a pretest survey to investigate how the verb-bias as a linguistic cue influence people's resolution to ambiguous pronouns. 30 healthy young adults from University of Shanghai for Science and Technology participated in the survey. We replace the nouns of each sentences with *X* and *Y* (for example, *X piping Y. Ta shengqi-le.* 'X blamed Y. She was angry.'), and ask the participants whether the pronoun refers to *X* or *Y*. On average, the object noun is preferred (74.5%).

We now apply the model to analyze the case shown in example sentences (4). Here are two possible worlds: w_1, where the owner was angry, and w_2, where the waiter was angry. Accordingly, the speaker's types include: the speaker knows that she is in w_1, say t_1, and she knows that she is in w_2, say t_2. From the survey analysis, the hearer's prior belief is biased towards t_2, that is, $p < \frac{1}{2}$. According to Proposition 1: If the speaker is of t_1, she will utter *The owner was angry*; if the speaker is of t_2, she will utter *He was angry*. The hearer, after receiving the message *He was angry*, will interpret it into t_2, namely, assigning the referent *the waiter* to the pronoun *he*.

The prediction of our analysis on ambiguous pronoun resolution is consistent with the results of our EEG experiment, which will be illustrated in the next section.

3 The Experiment

3.1 Methods

10 healthy adults from University of Shanghai for Science and Technology participated in the EEG experiment. All participants are native speaker of Chinese and right-handed. We excluded one participant due to significantly low accuracy rate of unambiguous pronoun resolution results. Therefore, all data analyses are based on 9 healthy adults.

We first construct 200 pairs of meaningful sentences in Chinese as described in Section 2.3. The sentences include both unambiguous pronouns and ambiguous pronouns. We not only construct sentences with the syntactic structure of $S+V+O$, (e.g. sentences (3) and (4)), but also construct pairs of sentences with unique syntactic structures of Chinese, N_1+N_2+V. Consider, for example, the following sentences:

(5) Laoshi he yanjiuyuan yuehui. Ta xiao-le.
teacher and researcher date. He smile ASP.
'The teacher dated the researcher. He smiled.'

The sentences are displayed on a computer screen. For each experimental trial, a fixation cross (500ms) shows first, and then three stimulus events follow (each 3000 ms). The first stimulus event is the presentation of a sentence containing two nouns (e.g., *Qizi lanzhu zhangfu.* 'The wife stopped the husband.'). The second stimulus event is the presentation of a sentence containing a pronoun (e.g., *Ta ku-le.* 'She cried.'). The last stimulus event is a question for participant to choose whether the pronoun refers to the noun on he left (e.g. *qizi* 'the wife') or the noun on the right (e.g. *zhangfu* 'the husband'). Participants are given up to 5500ms to respond, and their responses are recorded by tapping a certain key on the keyboard (key *z* as choosing the left noun, and key *m* for the right noun). All 200 pairs of sentences of either unambiguous classes or ambiguous classes are pseudo-randomly distributed. The experimental procedure is illustrated in Fig. 2.

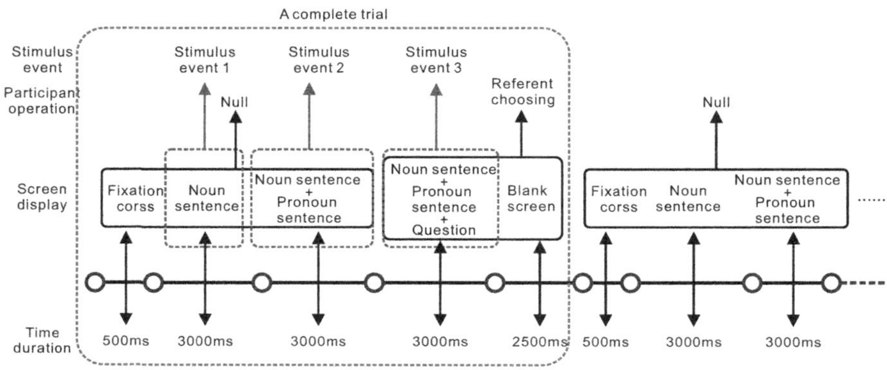

Fig. 2. The experiment procedure

EEG was recorded by Emotiv Xavier SDK at a sampling rate of 128 Hz using 14 Cu electrodes, placing according to international 10-20 system. The frequency bands included in EEG signals are as follows (see [27]): Theta (4-8 Hz), Alpha (8-12.5 Hz), Beta(12.5-28 Hz), Gamma (30-40 Hz). The schematic representation of the 14-channel positions is illustrated in Fig. 3(a).

3.2 Results and Discussion

We evaluate the response accuracy for the unambiguous pronoun resolution. The performance was highly accurate: M=95.8%. These results confirm the efficacy of gender information as useful for pronoun resolution. To evaluate the uses of verb-bias information, we analyze the proportion of participants assigning the referent to preferred object noun in the cases of ambiguous pronoun resolution. As predicted in the pretest survey (see Section 2.3), participants prefer choosing the object as the referent of pronoun (M=69%). We also analyze the cases of ambiguous pronoun resolution with a sentence structure of N_1+N_2+V, and participants show preference to the second noun as the referent of the pronoun (M=62%).

We investigate reaction times of pronoun resolution of various classes. We find that there are significant differences in reaction times between ambiguous and unambiguous pronoun resolution in both sentence structures, namely $S+V+O$ and N_1+N_2+V. More specifically, we report significant differences in reaction times collected from all 9 participants while resolving the following three groups of comparisons: Firstly, comparing unambiguous pronoun identified directly by both gender-biased nouns (e.g., sentences (3)) relative to ambiguous pronoun with the structure of $S+V+O$ (e.g., sentences (4)) shows a p value of t-test much smaller than 0.01(p=0.000); secondly, unambiguous pronoun identified indirectly by a gender-biased noun and a gender-neural noun (e.g., sentences (6)) versus ambiguous pronoun with the structure of $S+V+O$ (e.g., sentences (4)) (p=0.000); thirdly, unambiguous pronoun identified directly by gender-biased nouns (e.g., sentences (8)) versus ambiguous pronoun with the structure of N_1+N_2+V (e.g., sentences (5)) (p=0.000).

(6) Kuaiji guli nüer. Ta xiao-le.
accountant encourage daughter. he smile ASP.
'The accountant encouraged the daughter. He smiled.'

(7) Nüyanyuan he zhangfu zhengchao. Ta juezui-le.
actress and husband quarrel. she pout ASP.
'The actress quarrelled with the husband. She pouted.'

To investigate the neural correlates of pronoun resolution, we compare EEG data collected from 9 participants while processing unambiguous pronouns relative to the data of ambiguous pronoun resolution. More specifically, we report significant difference in EEG data collected from all 9 participants while resolving two groups of comparisons. Firstly, for EEG data collected from channel P8 for frequency bands Theta and Alpha, significant differences (p=0.005 for both frequency bands in signrank test) have been reported comparing resolution of unambiguous pronoun identified directly by gender-biased nouns (e.g., sentences (8)) relative to resolution ambiguous pronoun with the structure of N_1+N_2+V (e.g., sentences (5)). This result is illustrated as box plots in Fig. 3(c)-(d), where the significant difference from channel P8 is highlighted in red. Secondly, for EEG data collected from channel O1 for frequency bands Beta, a significant difference (p=0.006 in signrank test) has been reported comparing resolution of unambiguous pronoun identified directly by both gender-biased nouns (e.g., sentences (3)) relative to resolution ambiguous pronoun with the structure of $S+V+O$ (e.g., sentences (4)). This result is illustrated as box plots in Fig. 3(e)-(f), where the significant difference from channel O1 is highlighted in red.

The EEG signals from channels P8 and O1 are correlated with the recruitment of right inferior parietal cortex and left occipital cortex (see [28]). Right inferior parietal cortex is a region that has been correlated to an integration of of probabilistic assessment and evaluation of expected utility (see [20, 18]). Activation in occipital cortex has been reported as an increase of brain workload of the entire sentence reading (see [18]), which can be explained as correlated to increased demands of language processing.

4 Conclusions and Outlook

In this paper, we construct a game-theoretic model for ambiguous pronoun resolution. The behavioural findings suggest that people use gender information of nouns for pro-

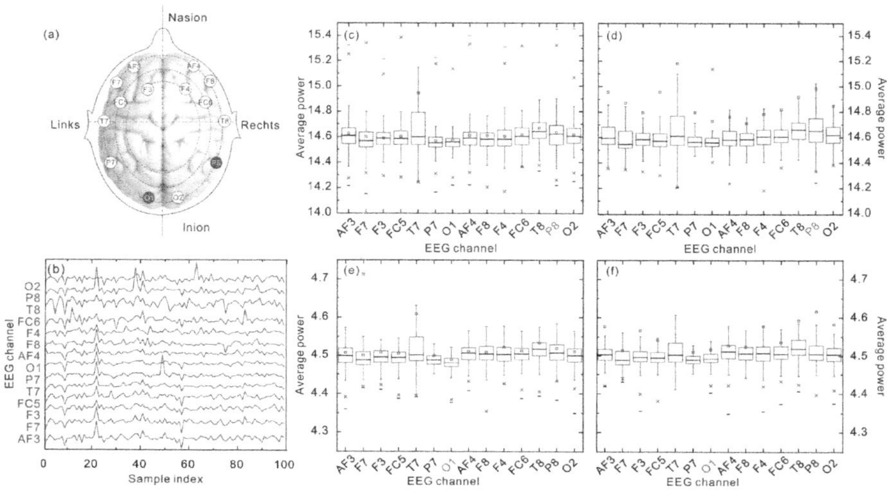

Fig. 3. EEG recording, localization of the electrodes position and signal difference comparing unambiguous to ambiguous pronoun processing. (a) Schematic representation of the 14 electrodes, and signals from channel P8 and O1 highlighted in red show significant difference between unambiguous and ambiguous pronoun processing. (b) EEG signals from 14 channels of a single participant during pronoun processing of 100 pairs of sentences, and signals from channel P8 and O1 are highlighted in red. (c)-(d) Box plots of EEG data of frequency Theta from 14 channels of 9 participants while processing unambiguous and ambiguous pronouns of $S+V+O$ and a significant difference shown in channel P8 is highlighted in red. (e)-(f) Box plots of EEG data of frequency Beta from 14 channels of 9 participants while processing unambiguous and ambiguous pronouns of N_1+N_2+V, and a significant difference shown in channel O1 is highlighted in red.

noun resolution with high accuracy. When gender is not informative, people use verb-bias information to resolve ambiguous pronouns. The experimental results are consistent with the assumptions and predictions of our model in three ways. Firstly, people spending more time in ambiguous pronouns than unambiguous pronouns is consistent with the assumption of our model that ambiguous pronoun resolution involves more complicated and time-consuming cognitive procedures, namely, decision-making. Secondly, people's consistent preference over object nouns in ambiguous pronoun resolution revealed in both pretest surveys and experiments is consistent with the prediction of our model that the solution to the game of ambiguous pronoun resolution is dependent on hearer's prior belief in speaker's types. Thirdly, significant differences shown in EEG channel P8 (right inferior parietal cortex) and O1 (left occipital cortex) comparing ambiguous to unambiguous pronoun processing implicates that the evaluation of probabilistic expected utilities are involved and accordingly increased demands of sentence processing are also involved, which is consistent with the assumption of our model.

The experimental results shown in this paper offers evidence that the epistemic processing of ambiguous pronoun resolution involves not only the core language network but also the network of strategic decision-making. These results coincide with the main assumption of the game-theoretic model: agents are making strategic decisions according to their belief in expected utilities. However, due to the limitation of EEG settings, it is extremely difficult to collect as well as analyze the neural data from both a speaker and a hearer at the same time during a conversation. To perform a closer test to the intersubjectivity of the agents assumed in the game model, the analysis can be further developed in the following two ways. First, to compare the predictions of the current model with corpus study results. A corpus study may provide with sentences in a certain context, and it may help to distinguish a speaker's intention from a hearer's interpretation. Second, to adopt psycholinguistic skills in the design of pretest. The psycholinguistic skills may help to separate the role of the speaker from that of the hearer by setting the pretest and the experiments apart.

References

1. Garvey, C., Caramazza, A.: Factors influencing assignment of pronoun antecedents*. Cognition **3**(3), 227–243 (1975)
2. Crawley, R., Stevenson, R., Kleinman, D.: The use of heuristic strategies in the interpretation of pronouns. Journal of Psycholinguistic Research **19**(4), 245–264 (1990)
3. Almor, A.: Noun-phrase anaphors and focus: The informational load hypothesis. Psychological Review **106**(4), 748–765 (1999)
4. Garnham, A.: Models of processing: Discourse. WIREs Cognitive Science **1**(6), 845–853 (2010)
5. Grice, P.: Logic and conversation. In: Cole, P., Morgan, J. (eds.) Syntax and Semantics, pp. 41–58. Academic Press, New York (1975)
6. Lewis, D.: Convention. Harvard University Press, Harvard (1969)
7. Parikh, P.: Communication and strategic inference. Linguistics and Philosophy **14**, 473–531(1991)
8. Parikh, P.: The Use of language. CSLI Publications, Stanford, CA (2001)
9. Clark, R.: Meaningful games: Exploring language with game theory. The MIT Press, Massachusetts (2011)

10. Clark, R., Parikh, P.: Game theory and discourse anaphora. Journal of Logic, Language and Information **16**, 265–282 (2007)
11. Frank, M. C., Goodman, N. D.: Predicting pragmatic reasoning in language games. Science **336**, 998 (2012)
12. Franke, M.: Signal to act: Game theory in pragmatics. Ph.D. thesis, Institute for Logic, Language and Computation, University of Amsterdam (2009)
13. Franke, M., Jäger, G.: Pragmatic back-and-forth reasoning. In: Reda, S. P. (eds.) Pragmatics, Semantics, and the Case of Scalar Implicatures, pp. 170–200. Palgrave, London (2014)
14. Brochhagen, T.: Signalling under uncertainty: Interpretative alignment without a common prior. The British Journal for the Philosophy of Science, axx058. http://doi.org/10.1093/bjps/axx058
15. Burkhardt, P.: Inferential bridging relations reveal distinct neural mechanisms: Evidence from event-related brain potentials. Brain and Language **98**, 159–168 (2006)
16. Ledoux, K., Gordon, P. C., Camblin, C. C., Swaab, T. Y.: Coreference and lexical repetition: Mechanisms of discourse integration. Memory and Cognition **35**(4), 801–815 (2007)
17. Nieuwland, M. S., Petersson, K. M., Van Berkum, J. J. A.: On sense and reference: examining the functional neuroanatomy of referential processing. NeuroImage **37**(3), 993–1004 (2007)
18. McMillan, C. T., Clark, R., Gunawardena, D., Ryant, N., Grossman, M.: fMRI evidence for strategic decision-making during resolution of pronoun reference. Neuropsychologia **50**(5), 674–687 (2012)
19. Brodbeck, C., Pylkkänen, L.: Language in context: Characterizing the comprehension of referential expressions with MEG. NeuroImage **147**, 447–460 (2017)
20. Huettel, S. A., Song, A. W., McCarthy, G.: Decisions under uncertainty: Probabilistic context influences activation of prefrontal and parietal cortices. Journal of Neuroscience **25**(13), 3304–3311 (2005)
21. Vickery, T., Jiang, Y.: Inferior parietal lobule supports decision making under uncertainty in humans. Cerebral Cortex **19**(4), 916–925 (2009)
22. Levinson, S. C.: Pragmatics. Cambridge University Press, New York (1983)
23. Sperber, D., Wilson, D.: Relevance: Communication and cognition. 2nd edn. Blackwell, Oxford (1995, 2004)
24. Jäger, G.: Game-theoretical pragmatics. In: van Benthem, J., ter Meulen, A. (eds.) Handbook of Logic and Language, pp. 467–491. Elsevier, Amsterdam (2011)
25. van Rooij, R., Franke, M.: Optimality-theoretic and game-theoretic approaches to implicature. The Stanford Encyclopedia of Philosophy (2015) http://plato.stanford.edu/entries/implicatureoptimality- games/
26. Benz, A., Stevens, J.: Game-theoretic approaches to pragmatics. Annual Review of Linguistics **4**, 173–191(2018)
27. Stickel, C., Fink, J., Holzinger, A.: Enhancing universal access—EEG based learnability assessment. Lecture Notes in Computer Science LNC **4556**, 813–822 (2007)
28. Zheng W., Lu B.: Investigating critical frequency bands and channels for EEG-based emotion recognition with deep neural networks. IEEE Transactions on Autonomous Mental Development **7**(3), 162–175 (2015)

On the Factivity Problem of Epistemic Contextualism*

Zhaoqing Xu

Department of Philosophy, Sichuan University
xuzhaoqing@scu.edu.cn

Abstract. Epistemic contextualists try to deal with skepticism by introducing contextual factors into semantic analysis of knowledge attributions. However, many researchers point out that the solution of epistemic contextualism faces the problem of factivity: its basic proposition conflicts with the factivity of knowledge. The key of epistemic contextualists is to distinguish the skeptical context of skepticism from the daily context, so as to solve the problem of skepticism in the sense of resolving a paradox. Therefore, I will also analyze how epistemic contextualists could respond to the factivity problem from two different perspectives: the skeptical context and the daily context. In the skeptical context, epistemic contextualists can distinguish two different kinds of skepticism, so that they can respond to the factivity problems of one version by significantly modifying their basic assertions. In the daily context, epistemic contextualists can distinguish the low standard daily context from the high standard daily context, so that they can respond to the factivity problem without significantly modifying their basic assertions. If we can use whether one needs to modifying the basic assertions in response to factivity problem as a theoretical benchmark to compare different versions of epistemic contextualism, then we can safely say that daily contextualism is better than skeptical contextualism. Along the way of argument, I make heavy use of the formal language and formal models of epistemic logic, but my informal interpretation of epistemic operators are different from the standard ones. I hope the whole discussion will illustrate a new application of epistemic logic and a relevantly neglected way of doing formal epistemology.

Keywords: Epistemic contextualism, factivity problem, skeptical contextualism, daily contextualism, epistemic logic, formal epistemology

1 Introduction

Epistemic contextualists try to deal with skepticism by introducing contextual factors into semantic analysis of knowledge attributions (e.g., [7]). However, many researchers point out that the solution of epistemic contextualism faces the

* This chapter is in its final form and it is not submitted to publication anywhere else.

B. Liao and Y. N. Wáng (eds.), *Context, Conflict and Reasoning*,
Logic in Asia: Studia Logica Library,
https://doi.org/10.1007/978-981-15-7134-3_5

problem of factivity: its basic proposition conflicts with the factivity of knowledge (e.g., [2] among others). The key of epistemic contextualists is to distinguish the skeptical context from the daily context, so as to solve the problem of skepticism in the sense of resolving a paradox. Therefore, I will also analyze how epistemic contextualists could respond to the factivity problem from two different perspectives: the skeptical context and the daily context.

Before going to the details, I shall make it clear that I am not an epistemic contextualist. And my goal is not to provide a strong defense for epistemic contextualism, but only a weak defense that as a mainstream theory of knowledge in contemporary epistemology, epistemic contextualism can avoid self-contradiction. Along the way of argument, I make heavy use of the formal language and formal models of epistemic logic, but my informal interpretation of epistemic operators are different from the standard ones. I hope the whole discussion will illustrate a new application of epistemic logic and justify a relevantly neglected way of doing formal epistemology.

2 Preliminaries

Definition 1 *Let Δ be a set of propositional letters $\{p, q, r\}$ and A be a set of indexes $\{1, 2, c\}$. The epistemic formulas of epistemic language are recursively defined as follows, where $p \in \Delta$ and $i \in A$:*

$$\varphi ::= p \mid \neg\varphi \mid (\varphi \to \psi) \mid K_i\varphi \mid KW_i.$$

Within the epistemic language, we can write various formulae, such as K_1p, KW_2q and $K_c(K_1p \to q)$. In standard epistemic logic, the indexes $\{1, 2, c\}$ are usually interpreted as different subjects, and K_1p and KW_2q read as "subject 1 knows that p" and "subject 2 knows whether q", respectively. However, in this paper, we will interpret the indexes differently. In the following sections, we will read K_1p, KW_2q and $K_c(K_1p \to q)$ as "in context 1 the subject knows that p", "in context 2 the subject knows whether q" and "in context 2 the epistemic contextualists knows whether in context 1 the subject's knowledge that p is the case", respectively. [1]

Definition 2 *An epistemic model $\mathfrak{M}$ is a tuple $\langle W, R_i(i = 1, 2, c), V\rangle$, where: W is the set of epistemically possible situations; R_is are reflexive binary relations on W and $R_1 \subseteq R_2$; V is a valuation, which specifies whether each propositional letter is true or false at each epistemically possible situation.*

We assume that the relations R_i are reflexive to ensure the factivity principle of knowledge, and we assume that $R_1 \subseteq R_2$ to reflect that the subject in context 2 has more uncertainty than the subject in context 1. [2]

[1] To make the distinction between context and subject more clear, one could use two indexes for each epistemic operator and write something like K_2^c. But since we will only discuss about c's knowledge in context 2, it is not necessary to use more complex symbols.

[2] If we have negative introspection, things will be more interesting and more complicated, but I have no space for detailed discussion here.

Definition 3 *Let $\mathfrak{M}$ be an epistemic model and w be an epistemic situation in $\mathfrak{M}$. The truth of φ at $\mathfrak{M}, w$ (denoted as $\mathfrak{M}, w \vDash \varphi$) is defined recursively as follows (I omit the Boolean cases):*

- $\mathfrak{M}, w \vDash K_i\varphi$ *iff for every v satisfying wR_iv, $\mathfrak{M}, v \vDash \varphi$.*
- $\mathfrak{M}, w \vDash KW_i\varphi$ *iff for every v satisfying wR_iv, $\mathfrak{M}, v \vDash \varphi$, or for every v satisfying wR_iv, not $\mathfrak{M}, v \vDash \varphi$.*

Finally, we define the notion of validity of epistemic formulas.

Definition 4 *A formula φ is valid (denoted as $\vDash \varphi$) iff $\mathfrak{M}, w \vDash \varphi$ for any epistemic model $\mathfrak{M}$ and any possible situation w.*

It follows immediately from Definition 4 that, a formula φ is not valid iff there is an epistemic model $\mathfrak{M}$ and a possible situation w such that not $\mathfrak{M}, w \vDash \varphi$.

Definition 5 *The set of all valid epistemic formulas can be axiomatized as follows (denoted as CEL, abbreviates "contextualists' epistemic logic").*

- *1. All substitutional instances of propositional tautologies.*
- *2.* $K_i(\varphi \rightarrow \psi) \rightarrow (K_i\varphi \rightarrow K_i\psi)$.
- *3.* $KW_i\varphi \leftrightarrow (K_i\varphi \vee K_i\neg\varphi)$.
- *4.* $K_i\varphi \rightarrow \varphi$.
- *5.* $K_2\varphi \rightarrow K_1\varphi$.
- *6. Modus Ponens: From φ and $(\varphi \rightarrow \psi)$ derive ψ.*
- *7. Necessity: From φ derive $K_i\varphi$.*

Theorem 1 *The logic CEL is sound and complete with respect to the class of all epistemic models. That is, for any epistemic formula φ, φ is a theorem of CEL (denoted as $\vdash_{CEL}$) iff $\vDash \varphi$.*

Proving completeness is a complicated routine in contemporary logic. Because of the limitation of space, I will not show any details of the proof here. [3] What is important for next sections is that, it follows from the standard proof of Theorem 1 that, a set Σ of epistemic formulas is consistent iff there exists an epistemic model $\mathfrak{M}$ and a situation w such that $\mathfrak{M}, w \vDash \varphi$ for all $\varphi \in \Sigma$.

3 The Factivity Problem

Consider the following two contexts: one is daily context (e.g., when you are in the supermarket buying your food), the other is skeptical context (e.g., when your epistemology teacher asks whether you know you are not a brain in a vat).

[3] Note that KW_i is not an independent operator, and the rest of the logic is very similar to Holliday's KT in [6]. The semantics I used here is a very traditional one. In [12], I developed a new semantics very similar to Holliday's L -semantics (of course, inspired by [7] and a previous immature version of [6]). The full version of [12] never get published, but see [13] for a similar semantics.

The subject in the daily context knows that p (e.g., "I has two hands"), while the subject in the skeptical context does not know that p, despite that the objective truth value of p and the internal cognitive state of the subject do not differ. (Note: we are talking about the same subject in two different contexts.) But the problem here is that, if in the daily context the subject knows that p, then in the skeptical context the subject knows that in the daily context she knows that p, and she also knows that the truth of knowledge in the daily context (the factivity of knowledge). According to most epistemic contextualists (e.g., [7]), when an epistemic contextualist is discussing the problem of skepticism, she is in the skeptical context, and she usually takes herself to be the subject under discussion. However, with some background assumptions, this will lead to a contradiction.

If we denote the daily context as context 1, and the skeptical context as context 2), then we can use the epistemic language (as in definition 1 to formally introduce the factivity problem as follows:

- (1) $K_1p \wedge \neg K_2p$:
  In the daily context the subject S knows that p , but in the skeptical context she doesn't know that p.
- (2) $K_1p \to K_2K_1p$:
  If in the daily context S knows that p, then in the skeptical context S also knows that (in the daily context S knows that p).
- (3) $K_2(K_1p \to p)$:
  In the skeptical context S knows that knowledge in the daily context implies true.
- (4) $K_2(K_1p \to p) \to (K_2K_1p \to K_2p)$:
  Knowledge in the skeptical context is also closed under known implication.

It is easy to deduce $K_1p \to K_2p$ from (2) (3) (4), which contradicts (1). Theoretically, we only need to reject either one of (1) - (4) to restore consistency. But the philosophical difficulty is, which one should go?

At a first glance, it seems strange to reject (1), because it means that what we know in the daily context is what we know in the skeptical context. This removes the foundation of epistemic contextualism, because then the daily context and the skeptical context will make no difference of knowledge attributions. (3) and (4) are only *a priori* principles of knowledge. More specifically, the principle of factivity (i.e., knowledge implies truth) is the least controversial principle of knowledge. Even in the context of skepticism, we can know that the principle is true. (4) is usually called "the epistemic closure principle". Although there are some philosophical controversies about this principle, rejecting it will pay a great theoretical price. What's more, one initial purpose for epistemic contextualists to introduce contextual factors into semantics of knowledge attributions is to defend the epistemic principle. Therefore, the most suspicious one here seems to be (2), because it is not so easy to know what you know in the same context, let alone know what you know in the daily context when you are in the skeptical context. But rejecting (2) will also lead to a very strange conclusion: $K_1p \wedge \neg K_2K_1p$.

If the subject under discussion is the epistemic contextualist herself, then it is easy to deduce that, $(K_1p \wedge \neg K_2K_1p)$ is a truth that she cannot know in the skeptical context. According to the knowledge norm of assertion (c.f., [11]), if she does not know that K_1p in the skeptical context, she should not even assert (1). That is why the factivity problem is also called the "knowability problem" or "assertability problem" (c.f.,[5].)

Theoretically speaking, epistemic contextualists have four choices: one is to modify the basic claims of epistemic contextualism; another is to reject (or doubt) the factivity principle of knowledge; the third is to reject the epistemic closure principle; and the fourth is to reject the knowledge norm of assertion. These four approaches are being promoted. Readers who want to know the current situation are referred to the latest survey article [9]. My plan here is not to directly compare the advantages and disadvantages of various proposals, but to ask what theoretical consequences we will arrive from rejecting the weakest (2). This line of thought was inspired by Nicholas Rescher's unified dissolution of various paradoxes [8], namely, rejecting the weakest chain of reasoning to contradiction. As mentioned above, if the epistemic contextualist do not know that K_1p in the skeptical context, he should not even assert (1). So, the question now is: what does she know and what can she assert in the skeptical context?

4 Analysis in the Skeptical Context

In my opinion, epistemic contextualists can distinguish two kinds of skepticism (this is inspired by Linton Wang and Oliver Tai's distinction between skeptical conclusions in [10]): one is p but S doesn't know that p, the other is S doesn't know whether p. Although the former is not the literal meaning of (1), it can be easily deduced from (1) with factivity of knowledge. And that's the problem. If the subject discuss the problem in the skeptical context, she don't know K_1p, so she can't make such an assertion; because in the daily context, she may or may not have hands. However, if we understand skepticism in the second way, in the daily context, the subject knows whether she has hands, rather than that she has. Similarly, in the skeptical context, the subject does not know whether she has hands or not.

According to my analysis here, epistemic contextualists can replace the original version of factivity problem with the following statements (KW abbreviates "know whether"):

- (1') $KW_1p \wedge \neg KW_2p$:
  In the daily context S knows whether p, but in the skeptical context she does not know whether p.
- (2') $KW_1p \rightarrow KW_2KW_1p$:
  If in the daily context S knows whether p, then in the skeptical context S knows whether in the daily context S knows whether p.
- (3') $KW_2(KW_1p \rightarrow p)$:
  In the skeptical context S knows whether knowledge-whether in daily context implies truth.

- (4') $KW_2(KW_1p \to p) \to (KW_2KW_1p \to KW_2p)$:
  Knowledge-whether in the context of skepticism also satisfies the principle of epistemic closure.

(3') is obviously true, because knowing whether p does not imply that p, epistemic contextualists can also know this in the skeptical context. But this time, (4') is not valid. (This can be proved using the method of model theory semantics. For details, see Proposition 3 of the Appendix; see also [4].) This leads to the first solution to the factivity problem in the skeptical context. Epistemic contextualists can consistently assert that: KW_1p, $\neg KW_2p$, KW_2KW_1p, and $KW_2\neg KW_2p$. (For a proof of the consistency of these assertions, see Proposition 5 of the Appendix.)

In fact, epistemic contextualists can also consider the following variation (note that (1") is the same as (1')):

- (1") $KW_1p \wedge \neg KW_2p$:
  In the daily context S knows whether p, but in the skeptical context doesn't know whether p.
- (2") $KW_1p \to K_2KW_1p$:
  If in the daily context S knows whether p, then in the skeptical context S knows that in the daily context S knows whether p.
- (3") $K_2(KW_1p \to p)$;
  In the skeptical context S knows that if in the daily context she knows whether p , then p is the case.
- (4") $K_2(KW_1p \to p) \to (K_2KW_1p \to K_2p)$.
  Knowledge in the skeptical context also satisfies the epistemic closure principle.

This time, it is obvious that (3") does not hold. (For a proof, see Proposition 4 of the Appendix.) This leads to the second solution to the factivity problem. Epistemic contextualists can consistently assert that: KW_1p, $\neg KW_2p$, K_2KW_1p and $K_2\neg KW_2p$. (For a proof of the consistency of these assertions, see Proposition 6 of the Appendix.)

The first variation is to transform the version of "know that" directly into a version of "know whether". The second variation is a mixture of "know that" and "know whether". If one only cares about restoring consistency, both approaches are feasible. But if we introduce more comparative standards, we find that the second one is superior. For example, the second approach is more congenial with the knowledge norm of assertion. Because the knowledge norm of assertion itself is expressed with "know that".

However, the problem here is that we are only bypassing the original factivity problem. The reason why "know whether" does not have "the factivity problem" is "know whether" does not have "factivity". But the factivity problem with "know that" still exists. This prompts us to think about other solutions. A closer look reveals that, both the original factivity problem and my analysis above are based on two assumptions: first, the epistemic contextualist is automatically in the skeptical context when she is expressing her theory, especially when she is discussing skepticism; second, the epistemic contextualist is the subject under discussion, especially in the skeptical context. However, it is not obligatory for epistemic contextualists to accept both assumptions. What if the epistemic contextualist can express her theory in the daily context, or what if the epistemic contextualist can be distinct from the subject under discussion? Let's now turn to my second analysis.

5 Analysis in the Daily Context

Although most epistemic contextualists admit that when the problem of skepticism is raised, one is automatically in the context of skepticism. However, there are still a few epistemic contextualists who believe that epistemic contextualists can also express their theories and discuss skepticism in daily context (c.f., [1]). (Truth is in the hands of the minority? Perhaps!) Therefore, in this section, we analyze how epistemic contextualists could respond to the factivity problem from the perspective of the daily context.

Let's look at the first possible variation (note that (a) is the same as (1)):

- (a) $K_1p \wedge \neg K_2p$:
  In the daily context S knows that p, but in the skeptical context she don't know that p.
- (b) $K_1p \rightarrow K_1K_1p$:
  If S knows p in the daily context, in the skeptical context, S also knows she knows that P in the daily context.
- (c) $K_1(K_1p \rightarrow p)$:
  S knows that in the daily context knowledge in the daily context implies truth.
- (d) $K_1(K_1p \rightarrow p) \rightarrow (K_1K_1p \rightarrow K_1p)$:
  Knowledge in the daily context is closed under the known implication.

This time, there is no contradiction. If epistemic contextualists can express his theory and skepticism in the daily context, they can consistently assert that: K_1p, $\neg K_2p$, K_1K_1p, $K_1\neg K_2p$. (For a proof, see Proposition 7 of the Appendix.) Some readers might have a question that: if one doesn't enter into the skeptical context, how should she explain $\neg K_2p$ and $K_1\neg K_2p$? I used the theoretical tools of dynamic epistemic logic to give an explanation at somewhere else. The basic idea is that the daily knowledge is with presuppositions. When the presuppositions face the challenge of skepticism, we do not automatically enter into the skeptical context, but can do counterfactual thinking in the daily context: what if these presuppositions do not hold? In that case, some knowledge will be lost due to the withdrawal of presuppositions. So the fundamental assertion of epistemic contextualism can be expressed as: when these basic presuppositions are established, the subject knows that p; when these basic presuppositions are not established, the subject does not know that p. Relevant technical details can be found in [13].

Although I tend to agree with the above analysis, due to the technical complexities its persuasiveness may be reduced. In addition, opponents may also say that the context with presuppositions established and the context without are two different contexts; or, if one can distinguish the two situations of knowing and not knowing through different presuppositions in the daily context, why bothers to introduce the skeptical context in the first place? In a word, if one does not enter into the skeptical context, the doubts about how could she explain that $\neg K_2 p$ and $K_1 \neg K_2 p$ have not been completely dispelled. Or, even if one does not enter into the skeptical context, it seems that one needs to enter into the context represented by subscript 2.

In my opinion, in order to dispel this kind of doubt, epistemic contextualists can also have a second response. They could distinguish two kinds of daily context: the low standard context and the high standard one. Now K_1 represents the subject's knowledge in the low standard context, and K_2 represents the subject's knowledge in the high standard context. (Note: the informal interpretations of epistemic operators now are different from above sections.) And K_c represents the epistemic contextualist's knowledge in the high standard context. As mentioned earlier, the knowledge norm of assertion requires that if $\neg K_2 K_1 p$, then $K_1 p$ should not be asserted. It was assumed that the subject in the skeptical context is the epistemic contextualist, that is, $K_c = K_2$. But what about $K_c \neq K_2$?

Consider the following variation (again, (a') is the same as (1), only that now we denote the low standard daily context as context 1, and the high standard daily context as context 2)):

- (a') $K_1 p \wedge \neg K_2 p$:
  In the low standard daily context S knows that p, but in the high standard daily context she doesn't know that p.
- (b') $K_1 p \to K_c K_1 p$:
  If in the low standard daily context S knows that p, then in the high standard daily context the epistemic contextualist knows that in the low standard daily context "S" knows that p.
- (c') $K_c(K_1 p \to p)$:
  In the high standard daily context, the epistemic contextualist knows that S's knowledge in the low standard daily context implies truth.
- (d') $K_c(K_1 p \to p) \to (K_c K_1 p \to K_c p)$:
  The epistemic contextualist's knowledge in the high standard daily context is also closed under known implication.

This time, (a') - (d') obviously have no contradiction (and are structurally parallel to (a)-(d)). Therefore, epistemic contextualists can consistently assert that: $K_1 p$, $\neg K_2 p$, $K_c K_1 p$ and $K_c \neg K_2 p$.(For a proof, see Proposition 8 of the Appendix.)

But how could the epistemic contextualist have more knowledge than the subject in the high standard daily context? Here is one possible explanation: her cognitive state is better than the epistemic subject, even they are both in the high standard daily context. For example, consider the classical zebra case of Dretske [3]. An ordinary subject might not know wether the animal in front of her is a zebra or a cleverly disguised mule, but being a trained animal morphologist (usually a philosopher is not a trained animal morphologist, but let's suppose our epistemic contextualist is), the epistemic contextualist can affirm that that this animal is a zebra rather than a cleverly disguised mule, according to the specific shape of the animal (such as the shape of its ears, hooves, etc.). Therefore, even in the high standard daily context, the epistemic contextualist could have more knowledge than ordinary subject, as long as she has a better cognitive state than the subject. So she could also have K_cK_1p, even if it is the case that $\neg K_2K_1p$. If one compare the two solutions of this section, one can safely say that, the lower knowledge standard and the better cognitive state are equivalent with respect to the effect of knowledge attributions. Some might doubt why K_c and K_2 are knowledge in the daily context, rather than knowledge in the skeptical context. My quick reply is that, if otherwise, it is hard to imagine (let alone justify) how the epistemic contextualist could have a better cognitive state than the subject under her discussion.

6 Reply to an Objection

Objection: You adopt the techinical method of epistemic logic to analyze the problem, not only use the formal language, but also the formal semantics (I should add, but also the proof system), but epistemic logic, though extremely weak in many senses, is known to have all kinds of problems from the view point of philosophy. This direction is generally problematic.

Reply: Yes, directly using epistemic logic to deal with philosophical problems is generally prblematic for philosophical reasons. The reason why epistemic logic can directly be used to discuss here is that epistemic contextualists generally accept the factivity and epistemic closure principle of knowledge, and their validity (for their proofs, see Proposition 1 and 2 of the Appendix) abductively justifies the using of epistemic semantics. I could tell the story in the other way and claim that from some philosophical observations I find a "new" epistemic logic, which technically happens to be the same as some old epistemic logic. I intentionally not do that to emphasize the other direction of applying epistemic logic. After all, reinterpretating epistemic operators is a common practice in the study of epistemic logic. For example, when challenged by philosophers, the knowledge operator K_i is usually reinterpreted as representing "semantic information" or "implicit knowledge".

7 Conclusion

I have analyzed the factivity problem of epistemic contextualism from the perspective of the skeptical context and the daily context respectively. In the skeptical context, epistemic contextualists could distinguish two different kinds of skeptical conclusions, so as to respond to the factivity problem via a modification of their basic assertion (1). In the daily context, the epistemic contextualists could distinguish the low standard daily context and the high standard daily context, so that they could respond to the factivity problem without modifying their basic assertion. If we can use whether one needs to revise the basic assertion (1) in response to the factivity problem as the theoretical benchmark for comparing different versions of epistemic contextualism, then we can safely say that daily contextualism (epistemic contextualism in the daily context) is better than skeptical contextualism (epistemic contextualism in the skeptical context).

During the presentation of my analyses, I used not only the formal language of epistemic logic, but also its semantic models. I hope the whole discussion have illustrated that a relevantly neglected way (especially by the philosophy community) of doing formal epistemology is still be fruitful (or at least feasible).

Finally, I point out some issues for further research: (i) extend the analyses here to cover other structurally similar problems (e.g., the knowledge transitive problem in multi-agent epistemic logic, and the nesting problem of two-dimensional semantics); (ii) extend the epistemic language to cover group knowledge operators (e.g., distributive knowledge, collective knowledge, and common knowledge); (iii) compare CEL in Section 2 with other relevant systems (e.g.,[6]).

Acknowledgement

The work was supported by National Social Science Fund Of China (18CZX063) and Sichuan University(2018hhs-50;SCU8303). Different drafts were reported at various occasions: International Conference on Truth, Logic and Philosophy (Peking University, September 23-24, 2017); Fifth Academic Conference of Chinese Association of Epistemology (Central South University, October 19-20, 2019); Annual Conference of China Society for History of Foreign Philosophy and China Society for Modern Foreign Philosophy (Sichuan University, November 2-3, 2019); Annual Conference of Sichuan Socety of Philosophy and Sichuan Society of History of Marxist Philosophy (Bazhong Vocational and Technical College, November 8-10, 2019); Workshop on Logic and Paradox (Anhui University, November 30-December 1, 2019); Symposium on the 40th anniversary of Taiyuan conference and the New Trend of Contemporary Western Philosophy (Shanxi University, December 6-9, 2019). I would like to thank Yang Jianguo, Fang Huanhui, Cao Jianbo, Wei Yidong, Zhao Zhen, Wang Shuqing, Chen Jia, and Zhang Shun, and three anonymous referees of AWPL 2020 (Zhejiang University) for useful comments and suggestions.

Appendix: Proofs

First, we prove that the factivity of knowledge and the epistemic closure principle are valid in any epistemic model. To prove the validity of a formula, we can use *reductio ad absurdum*.

Proposition 1. $K_2(K_1p \to p)$ *is valid.*

Proof. Assume $K_2(K_1 p \to p)$ is not valid, then there exists an epistemic model $\mathfrak{M}$ and possible situation w such that $\mathfrak{M}, w \vDash \neg K_2(K_1 p \to p)$, then v satisfying $wR_2 v$, such that $\mathfrak{M}, v \vDash \neg(K_1 p \to p)$, which means $\mathfrak{M}, v \vDash K_1 p$ and $\mathfrak{M}, v \vDash \neg p$. From reflexivity of R_1 (that is, $vR_1 v$ and $\mathfrak{M}, v \vDash K_1 p$, it follows that $\mathfrak{M}, v \vDash p$; from $\mathfrak{M}, v \vDash \neg p$, it follows that not $\mathfrak{M}, v \vDash p$. A contradiction. Therefore, the assumption does not hold, $K_2(K_1 p \to p)$ is valid.

Proposition 2. *$K_2(K_1 p \to p) \to (K_2 K_1 p \to K_2 p)$ is valid.*

Proof. Assume $K_2(K_1 p \to p) \to (K_2 K_1 p \to K_2 p)$ is not valid, then there exists a model $\mathfrak{M}$ and a possible situation w such that $\mathfrak{M}, w \vDash \neg(K_2(K_1 p \to p) \to (K_2 K_1 p \to K_2 p))$, which means $\mathfrak{M}, w \vDash K_2(K_1 p \to p)$ and $\mathfrak{M}, w \vDash K_2 K_1 p$, but not $\mathfrak{M}, w \vDash K_2 p$. From the later, it follows that there exists a v satisfying $wR_2 v$, such that $\mathfrak{M}, w \vDash \neg p$, that is not $\mathfrak{M}, v \vDash p$. But from the former, it follows that $\mathfrak{M}, v \vDash, v| = K_1 p \to p$ and $\mathfrak{M}, v \vDash K_1 p$, then according to semantics of $\to$, $\mathfrak{M}, v \vDash p$. A contradiction. So the assumption does not hold, $K_2(K_1 p \to p) \to (K_2 K_1 p \to K_2 p)$ is valid.

Then, we prove the invalidity of two formulas in Section 4. To prove that a formula is not valid, we only need to give an epistemic model, such that the formula is false at a possible situation of the model.(We omit R_c when K_c is not relevant.)

Proposition 3. *$KW_2(KW_1 p \to p) \to (KW_2 KW_1 p \to KW_2 p)$ is not valid.*

Proof. Consider the model $\mathfrak{M} = \langle W, R_1, R_2, V \rangle$, where $W = \{w, v, u\}$, $R_1 = \{(w,w), (w,v), (v,v), (v,u), (u,u)\}$, $R_2 = \{(w,w), (w,v), (v,v), (v,w), (v,u), (u,u)\}$, $V(p) = \{v\}$. Then $\mathfrak{M}, w \vDash p$ and $\mathfrak{M}, v \vDash p$, so $\mathfrak{M}, w \vDash KW_1 p$ and $\mathfrak{M}, v \vDash KW_1 p$. Hence $\mathfrak{M}, w \vDash KW_1 p \to p$ and $\mathfrak{M}, v \vDash KW_1 p \to p$. It follows that $\mathfrak{M}, w \vDash KW_2 p$, $\mathfrak{M}, w \vDash KW_2 KW_1 p$ and $\mathfrak{M}, w \vDash KW_2(KW_1 p \to p)$. So, $\mathfrak{M}, w \vDash \neg(KW_2(KW_1 p \to p) \to (KW_2 KW_1 p \to KW_2 p))$. Therefore, $KW_2(KW_1 p \to p) \to (KW_2 KW_1 p \to KW_2 p)$ is not valid.

Proposition 4. *$K_2(KW_1 p \to p)$ is not valid.*

Proof. Consider the model $\mathfrak{M} = \langle W, R_1, R_2, V \rangle$, where $W = \{w, v\}$, $R_1 = \{(w,w), (v,v)\}$, $R_2 = \{(w,w), (w,v), (v,v)\}$, $V(p) = \{v\}$. Then $\mathfrak{M}, w \vDash \neg p$, so $\mathfrak{M}, w \vDash \neg KW_1 p$, hence $\mathfrak{M}, w \vDash \neg(KW_1 p \to p)$; thus $\mathfrak{M}, w \vDash \neg K_2(KW_1 p \to p)$. Therefore, $K_2(KW_1 p \to p)$ is not valid.

Finally, we prove that the four groups of formulas mentioned in the text are consistent. To prove that, we only need to construct an epistemic model, such that the group of formulas are all true at a certain possible situation of the model.

Proposition 5. *$KW_1 p$, $\neg KW_2 p$, $KW_2 KW_1 p$ and $KW_2 \neg KW_2 p$ are consistent.*

Proof. Consider the model $\mathfrak{M} = \langle W, R_1, R_2, V \rangle$, where $W = \{w, v\}$, $R_1 = \{(w,w), (v,v)\}$, $R_2 = \{(w,w), (w,v), (v,v)\}$, $V(p) = \{w\}$. Then $\mathfrak{M}, w \vDash KW_1 p$, $\mathfrak{M}, w \vDash \neg KW_2 p$, $\mathfrak{M}, w \vDash KW_2 KW_1 p$, and $\mathfrak{M}, w \vDash KW_2 \neg KW_2 p$. Therefore, $KW_1 p$, $\neg KW_2 p$, $KW_2 KW_1 p$ and $KW_2 \neg KW_2 p$ are consistent.

Proposition 6. *KW_1p, $\neg KW_2p$, K_2KW_1p and $K_2\neg KW_2p$ are consistent.*

Proof. Consider the model $\mathfrak{M} = \langle W, R_1, R_2, V\rangle$, where $W = \{w, v\}$, $R_1 = \{(w,w),(v,v)\}$, $R_2 = \{(w,w),(w,v),(v,v),(v,w)\}$, $V(p) = \{w\}$. Then $\mathfrak{M}, w \vDash KW_1p$, $\mathfrak{M}, w \vDash \neg KW_2p$, $\mathfrak{M}, w \vDash K_2KW_1p$ and $\mathfrak{M}, w \vDash K_2\neg KW_2p$. Therefore, KW_1p, $\neg KW_2p$, K_2KW_1p and $K_2\neg KW_2p$ are consistent.

Proposition 7. *K_1p, $\neg K_2p$, K_1K_1p and $K_1\neg K_2p$ are consistent.*

Proof. Consider the model $\mathfrak{M} = \langle W, R_1, R_2, V\rangle$, where $W = \{w, v\}$, $R_1 = \{(w,w),(v,v)\}$, $R_2 = \{(w,w),(w,v),(v,v)\}$, $V(p) = \{w\}$. Then $\mathfrak{M}, w \vDash K_1p$, $\mathfrak{M}, w \vDash \neg K_2p$, $\mathfrak{M}, w \vDash K_1K_1p$, and $\mathfrak{M}, w \vDash K_1\neg K_2p$. Therefore, K_1p, $\neg K_2p$, K_1K_1p and $K_1\neg K_2p$ are consistent.

Proposition 8. *K_1p, $\neg K_2p$, K_cK_1p and $K_c\neg K_2p$ are consistent.*

Proof. Consider the model $\mathfrak{M} = \langle W, R_1, R_2, R_c, V\rangle$, where $W = \{w, v\}$, $R_1 = \{(w,w),(v,v)\} = R_c$, $R_2 = \{(w,w),(w,v),(v,v)\}$, $V(p) = \{w\}$. Then $\mathfrak{M}, w \vDash K_1p$, $\mathfrak{M}, w \vDash \neg K_2p$, $\mathfrak{M}, w \vDash K_cK_1p$, and $\mathfrak{M}, w \vDash K_c\neg K_2p$. Therefore, K_1p, $\neg K_2p$, K_cK_1p and $K_c\neg K_2p$ are consistent.

References

1. Barke, A.: Epistemic contextualism. Erkenntnis 61(2), 353–373 (2004)
2. Baumann, P.: Factivity and contextualism. Analysis 70(1), 82–89 (2010)
3. Dretske, F.: Epistemic operators. Journal of Philosophy 67, 1007–1023 (1970)
4. Fan, J., Wang, Y., van Ditmarsch, H.: Contingency and knowing whether. The Review of Symbolic Logic 8(1), 75–107 (2015)
5. Freitag, W.: On the knowability of epistemic contextualism: A reply to m. montminy and w. skolits. Episteme 12(3), 335–342 (2015)
6. Holliday, Wesley, H.: Epistemic closure and epistemic logic i: Relevant alternatives and subjunctivism. Journal of Philosophical Logic 44(1), 1–62 (2015)
7. Lewis, D.: Elusive knowledge. Australasian Journal of Philosophy 74(4), 549–567 (1996)
8. Rescher, N.: Paradoxes: their roots, range, and resolution. Open Court, (2001)
9. Stefano, L., Nicla, V.: Knowledge in context: The factivity principle and its epistemological consequences. Kairos: Journal of Philosophy and Science 20(1), 12–42 (2018)
10. Wang, L., Tai, O.: Skeptical conclusions. Erkenntnis 72(2), 177–204 (2010)
11. Williamson, T.: Knowledge and its Limits. Oxford University Press (2000)
12. Xu, Z.: Capturing lewis's "elusive knowledge". In: van Ditmarsch, H., Lang, J., Ju, S. (eds.) Logic, Rationality, and Interaction. pp. 400–401. Springer Berlin Heidelberg, Berlin, Heidelberg (2011)
13. Xu, Z., Chen, B.: Epistemic logic with evidence and relevant alternatives. In: van Ditmarsch, H., Sandu, G. (eds.) Jaakko Hintikka on Knowledge and Game-Theoretical Semantics, pp. 535–555. Springer International Publishing, Cham (2018)

Dynamic Epistemic Logic of Faith Diffusion in Cultural Circles*

Junli Jiang

Institute of Logic and Intelligence, Southwest University, Chongqing 400715, China
jiangjl2014@swu.edu.cn

Abstract. How does faith diffuse in a social network that the same person may belong to different cultural circles and with different threshold values to their influence? In this article we refer to Morrison and Naumov's group conformity model with thresholds as the cultural circle based threshold models studying the faith diffusion on social networks. Cultural circle based threshold model consists of a network graph with agents in different cultural circles and the corresponding cultural circle threshold values which regulate the diffusion. We show that the recently proposed "propositional opinion propagation model" based model modelling an agent's conformity to different social groups and the threshold model modelling an agent's conformity to the society as whole are two special cases of the cultural circle based threshold model. We construct a dynamic epistemic logic to describe the diffusion dynamics and the epistemic properties of the influence relation in cultural circle based threshold model.

Keywords: cultural circle · threshold model· faith diffusion· social network · dynamic epistemic logic.

1 Introduction

A person may belong to different cultural circles at the same time, such as classmates circle, colleagues circle, community neighborhood group, and so on. Different cultural circles have different influences. Then under what circumstances can cultural circles influence you? Do you have the same tolerance for different cultural circle based influences? If you don't know for sure that who are in your cultural circles or their action, then will they affect you? This paper focuses on answering these questions through dynamic epistemic logic.

Several logical frameworks for reasoning about diffusion in social networks have been studied before. Seligman et al. [10, 11] and Liu et al. [7] proposed logics for capturing properties of Facebook and belief change in epistemic social networks based on modal languages. Christoff et al. [3] proposed minimal threshold influence logic that uses modal language to capture dynamic of diffusion in a

* Supported by the Key Research Funds for the Key Liberal Science Research Base of Chongqing under research no.16SKB036. This chapter is in its final form and it is not submitted to publication anywhere else.

B. Liao and Y. N. Wáng (eds.), *Context, Conflict and Reasoning*,
Logic in Asia: Studia Logica Library,
https://doi.org/10.1007/978-981-15-7134-3_6

threshold model and gave a complete axiomatization of this logic. Baltag et al. [2] discussed dynamic epistemic logics for informed update and prediction update. Xiong et al. [12] formalised the followership in networks that agents following or unfollowing each other dynamically through hybrid logic. Naumov et al. [9] used Armstrong's axioms to describe influence in social networks, Morrison and Naumov further developed this approach in paper [8] where they introduced Partition axiom to study properties of influence relation for a given network topology in a setting of propositional opinion diffusion model.

This paper has three goals. Our first goal is to propose logical frameworks for reasoning about cultural circle based threshold models and their dynamics. Our second goal is to investigate how cultural circle can affect faith diffusion. Our third goal is to investigate how the agents' knowledge affects such dynamics. After informally recalling the standard threshold model and its limitations in capturing the influence of cultural circles in Section 2, a cultural circle based threshold model for modeling cultural circle influence on agents is introduced in Section 3. Instead of modelling the influences of cultural circles when agent adopt a behavior, we focus in Section 4 on agents who adopt when they know what kind of information. In this section we introduce epistemic threshold models. These models come equipped with a specific knowledge-dependent update procedure, called "knowledge based update", where agents must possess sufficient information about their surroundings before they adopt. We concluded the section by extending the logic to a sound and complete dynamic epistemic logic for the epistemic circle based threshold models and the knowledge based update procedure. Some conclusions are given in Section 5.

2 Threshold Models and Cultural Circles

The most commonly studied model of diffusion in social networks is the threshold model [6]. The threshold model with the assumption that there is a "seed" set B of agents who already adopted (or infected) assigns threshold values to all agents in the network and a directed edge represents possible influence of one agent on another. If the sum of incoming edges from agents that already adopted the faith (or infected with the virus) is at least as much as agent's threshold value, then the agent also adopts the faith(or infected). Thus threshold models and their updates could be used to interpret two kinds of social phenomena. In one social phenomenon that agents have no autonomy, and their behavior is forced upon them by their environment, e.g. pandemics. This model could be used in epidemiology. In another social phenomenon that agents may choose action aiming towards coordination with their environment, e.g. conformity.

An example of such standard threshold model on social network is partly depicted in Fig. 1 where the threshold value of agent$a$ is 0.7. Suppose agent e in this example is given a new faith and it have adopted it, then the agent puts peer pressure $\frac{1}{4}$ on agent a to adopt the new faith. Since $\frac{1}{4}$ is less than its threshold value 0.7, this pressure alone will not lead to agent a adopting the faith. However, if agents b and c are given this new faith and they have adopted it, then the total peer pressure on agent a becomes $\frac{3}{4}$ which is greater than its threshold value 0.7. Thus, after agents b, c and e adopting the faith, they will influence agent a to do the same. In a larger social network, the process of the adoption might continue with more and more agents putting peer pressure on others and, as a result, more and more agents adopting the faith. This process is called faith diffusion in social networks.

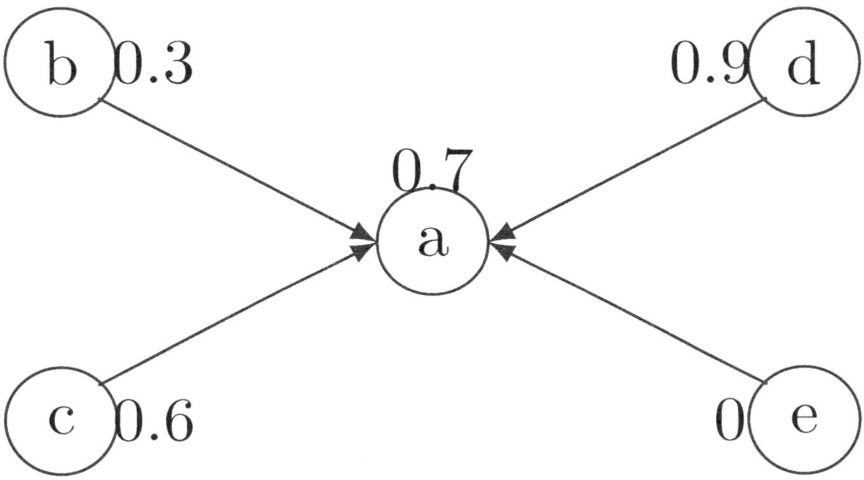

Fig. 1. Threshold model

A significant limitation of the standard threshold model comes from the fact that it treats peer pressure equally no matter which cultural circles it comes from. However the fact is that once the peer pressure come from one cultural circle, it will bring more push power than its even distribution among cultural circles to agent. Consider the following variant example of [8] , assume an agent a has about the same number of two types of peers: co-workers and relatives. If the person learns that 80% of co-workers travel abroad, then she would feel peer pressure to confirm to the group's norms and to travel abroad even if no relatives travel abroad. Similarly, if agent a learns that 80% of her relatives travel abroad, then she would feel pressured into traveling abroad even if most of her co-workers don't. At the same time, it is plausible that the agent might experience much less pressure to travel abroad if her 40% co-workers and 40% relatives travel abroad.

Morrison and Naumov [8] use group conformity model to overcome this limitation. They use a Boolean function which can be represented as disjunctions of conjunctions of atomic Boolean variables to model influence on an agent, and each disjunct in the disjunction can be viewed as a conformity group of the agent.

For the above example with co-workers and relatives, the Boolean function of agent a could be $(b \wedge c) \vee (d \wedge e)$. Disjuncts $(b \wedge c)$ and $(d \wedge e)$ correspond to conformity groups$\{b, c\}$ and $\{d, e\}$. If all agents in at least one conformity group adopt the product, then agent a also adopts the product. This model can handle this extreme case well, but leave the more general situations that a certain percentage of the population in one conformity group is enough to make agent adopts the product behind. In this article we develop their idea by the notion of cultural circle with threshold to handle the more general situations as in the above example.

3 Cultural Circle Based Threshold Models and Information Diffusion

A social network consist of a set of agents and a social relationship, it can also be seen as a graph where nodes represent agents and edges represent a binary social relationship among them. It is assumed throughout this paper that both social relationships and cultural circles on a social network stay constant under faith updates. We consider the "simplest" possible network structures: the networks are finite, symmetric, irreflexive and serial, which correspond to imposing constraints on social networks: finite, connected, directed, and without self-loop. The constraints of symmetry and irreflexivity could easily be relaxed in the following initial definition of cultural circle based threshold models to generalize the logic to different types of social relationships.

Definition 1 (Social Network With Cultural Circles). *A social network with cultural circles is a tuple* $CSN = (A, S, N, \{C_a\}_{a \in A})$ *where*

1. *A is a non-empty finite set of agents.*
2. *The function $S : A \longrightarrow \mathcal{P}(A)$ assigns a social circle Sa to each $a \in A$ such that*
   - *(Irreflexivity)$a \notin Sa$,*
   - *(Symmetry)$b \in Sa \Leftrightarrow a \in Sb$,*
   - *(Seriality)$Sa \neq \emptyset$.*
3. *The function $N : A \longrightarrow \mathcal{N}$ assigns a natural number to each $a \in A$ representing the number (Na) of cultural circles a has, and $I_a = \{i | 1 \leq i \leq Na\}$.*
4. *The function $C_a : I_a \longrightarrow \mathcal{P}(Sa)$ assigns a cultural circle C_a^i to each $i \in I_a$ such that*
   - *$C_a = \{C_a^i | i \in I_a\}$ where $C_a^i \subseteq Sa$ is agent a's cultural cirle,*
   - *$\bigcup_{i=1}^{Na} C_a^i = Sa$.*

In this definition $\forall a \in A$, $i, j \in I_a$, $(C_a^i \cap C_a^j)$ could be a nonempty set, since you sometimes do be a friend and a co-worker of the same person, and it is not natural or necessary to constrain the intersection to be a empty set. We use C_a^i to generally represent a's any cultural circles when there is not need to identify them. Sa is the set of all agents in set A that are connected by a directed edge to agent a. Here different edges may represent different social relationships. When all edges are represent the same one social relationship, then agent a only has one cultural circle.

Definition 2 (Cultural Circle Based Threshold Model). *Cultural circle based threshold model* $M = (A, B, S, N, \{C_a\}_{a \in A}, \theta)$ *where set* $B \subset A$ *is the set of agents who have already adopt the new faith and* $\forall a \in A$ *the function* $\theta : C_a \longrightarrow [0, 1]$ *assigns a threshold* $\theta_{C_a^i}$ *to a's each cultural circle* C_a^i

The above cultural circle based threshold model will degrade into the standard threshold model when every agent has only one cultural circle or just see Sa as a whole without considering cultural circle. Obviously, the above cultural circle based threshold model will degrade into the group conformity model when every agent assigns 1 as the only threshold value to their each cultural circle.

The following conception F_a is the set of all possible agent sets within each cultural circle that can make agent a adopts what they have adopted.

Definition 3 (Upward Closed Set of Influence within Cultural Circle). *For* $a \in A$, *set* $F_a = \{X \mid \forall C_a^i \in C_a, X \subseteq C_a^i, |X| \geq \theta_{C_a^i}|C_a^i|\}$ *is the union upward closed set of a's influence sets within its each cultural circle.*

It is helpful to use an example to explain the conception in the above definition: Consider a simplest social network in which an agent a belongs to two cultural circles $C_a = \{C_a^1, C_a^2\}$ where $C_a^1 = \{b, c\}$, $C_a^2 = \{d, e\}$ and suppose $\theta_{C_a^1} = 0.5$, $\theta_{C_a^2} = 1$. Therefore the upward closed influence sets in cultural circle C_a^1 be the set$F_a^1 = \{\{b\}, \{c\}, \{b, c\}\}$ and the upward closed influence set in cultural circle C_a^2 be the set $F_a^2 = \{\{d, e\}\}$, $F_a = F_a^1 \cup F_a^2$ which means $F_a = \{\{b\}, \{c\}, \{b, c\}, \{d, e\}\}$.

Now we arrive at the place to give out the key definition based on the above conception. The following conception Y_a is the set of all possible agent sets within a's social circle that can make agent a adopts what they have adopted.

Definition 4 (The Smallest Upward Closed Set of Influence On Social Network). *For* $a \in A$, *set* $Y_a = \{Y \mid Y \in \mathcal{P}(Sa), X \in F_a, X \subseteq Y\}$ *is the upward closed set of the sets in which agents can influence a on social network.*

The word "*smallest*" means that we only consider the agents that direct to a (in Sa), not all agents (in A) on the social network. As you can see that $Y_a \subseteq \mathcal{P}(Sa)$ and equal to the upward closed set of set Sa's subset on F_a. We can use the above example to explain this idea more details. Since $Sa = \cup C_a = \{b, c, d, e\}$ and $F_a = \{\{b\}, \{c\}, \{b, c\}, \{d, e\}\}$, we have $Y_a = F_a \cup \{\{b, d\}, \{b, e\}, \{b, d, e\}\} \cup \{\{c, d\}, \{c, e\}, \{c, d, e\}\} \cup \{\{b, c, d\}, \{b, c, e\}, \{b, c, d, e\}\}$, so every element in Y_a can have potential influence to agent a. Y_a is the set of a's all possible influence sets across her cultural circles on social networks.

The idea of above two definitions comes from [8] in which Morrison and Naumov introduce the notion of upward closed family of subsets. When every cultural circle threshold is equal to 1 for each agent, then above two definitions will respectively degrade into the set of cultural circle and the following notion similar to Definition3 in [8]:

Definition 5 (Upward Closed Set On Cultural Circles). *For $a \in A$, set Y_a is upward closed set on a's cultural circles if $\forall C_a^i \in C_a$ such that $C_a^i \in Y_a$ and for any subset $S_a^j \subseteq Sa$ if $\exists C_a^i \in C_a and C_a^i \subseteq S_a^j$, then $S_a^j \in Y_a$.*

It is assumed throughout this paper that both the cultural circles on network structure and the adoption thresholds stay constant under updates. We use $Sa(B) = \{b \mid b \in B \text{ and } b \in Sa\}$ to represent the intersection of set Sa and set B. It is natural to assume that once an agent adopted a faith she or he would always adopt it. Therefore, the spread of the faith (i.e., the extension of B) at ensuing time steps may be calculated using the fixed thresholds and cultural circles as follows:

Definition 6 (Faith Update).

$$B_n = \begin{cases} B & \textit{if } n = 0, \\ B_{n-1} \cup \{a \in A \mid Sa(B_{n-1}) \in Y_a\} & \textit{if } n > 0. \end{cases}$$

The new set of agents B_1 who have adopted does include those agents in set B and agents who have enough neighbors that have adopted the faith already in Y_a. We can obtain a unique sequence of cultural circle based threshold model by repeatedly applying this update rule in present cultural circle based threshold model.

Definition 7 (Diffusion Sequence). *Let $M = (A, B, S, N, \{C_a\}_{a \in A}, \theta)$ be a Cultural circle based threshold model. The diffusion sequence $\mathcal{L}_M$ is the sequence of threshold models $\langle M_0, M_1, \ldots, M_n, M_{n+1} \rangle$ such that for any $n \in \mathcal{N}$, $M_n = (A, B_n, S, N, \{C_a\}_{a \in A}, \theta)$ where B_n can be calculated by the above definition 6. There exists a natural number $n < |A|$ such that this diffusion process always reaches a fixed point $M_n = M_{n+1}$.*

Since it is assumed throughout this paper that both the network structure and the adoption threshold stay constant under updates, the diffusion of the new faith will stop after reaching the fixed point. the longest update time is $(|A| - 1)$ by considering the slowest possible diffusion scenario, i.e. where $|B_0|$=1 and only one agent adopts per round.

4 Logics for Faith Diffusion

By the definition of update on cultural circle based threshold models, agents are always influenced by the actual behavior of their direct neighbors, which means that update requires agents act in accordance with the facts of others' behavior. To express how social faith changes according to a round of faith update, we use the language DL which extends the propositional language with a dynamic modality [adopt] to build formulas $[adopt]\phi$ ("after a round of faith update, ϕ is the case").

Definition 8 (Dynamic Language DL). *Formulas in DL are given recursively as following:*

$\phi := \beta a \mid Sba \mid C_a^i ba \mid \neg\phi \mid \phi \wedge \phi \mid [adopt]\phi$

The language is interpreted over cultural circle based threshold models, using the behavior set and the social network to determine the extension of the atomic formulas. The [adopt] modality is interpreted as is standard in dynamic epistemic logic [4].

Definition 9 (Semantic of DL**).** *Given a cultural circle based threshold model* $M = (A, S, B, \{C_a\}_{a\in A}, \theta)$ *and the formulas in DL:*

$M \models \beta a$ *iff* $a \in B$
$M \models Sba$ *iff* $b \in Sa$
$M \models C_a^i ba$ *iff there is a* $i \in I_a$ *such that* $b \in C_a^i$
$M \models \neg\phi$ *iff* $M \not\models \phi$
$M \models \phi \wedge \psi$ *iff* $M \models \phi$ *and* $M \models \psi$
$M \models [adopt]\phi$ *iff* $M_1 \models \phi$, *where* M_1 *is the faith update of* M.

Intuitively, $[adopt]\phi$ is true in a given model if and only if ϕ is true in the model after a given change occurs. Here, this change means that all agents simultaneously update their faith according to Definition 6 of faith update. We can build the following axiom system via the standard method of reduction rules from dynamic epistemic logic [4]. It is worth to note that the following propositional formula $\beta_{b\in G}$ expresses that there is a set of agents who have adopted the faith in F_a and can give a enough peer pressure:

$$\beta_{b\in G} := \bigvee_{G\in F_a} \bigwedge_{b\in G} \beta b \tag{1}$$

Hence, the following DL formula states that agent a will have adopted the faith after the update just in case she had already adopted it before the update or she currently has a set in F_a that the agents in which have already adopted it.

$$[adopt]\beta a \leftrightarrow \beta a \vee \beta_{b\in G} \tag{2}$$

We can obtain an axiomatization of the logic for cultural circle based threshold models and their update dynamics like what [2, 3] have done by using the standard method of reduction rules from dynamic epistemic logic [4]. We wouldn't discuss it here.

However in many situation agents may do not know this kind of information about their social circle, which means that they can't coordinate their action in according with other agents in their social circle. Hence, To make sure the above Definition 6 of faith update do happen, it is requires that agents should first know it. For example, consider the situation in which you will buy a new production if half of your friends have already bought it, you don't know whether the actual ratio have reached or not, although all of them have already bought it. So there are two necessary information for you to buy the new production: whether the actual ratio have reached your threshold and whether you know it.

To accommodate this shortcoming in cultural circle based threshold models, we add an epistemic dimension to it and define a refined adoption policy where agents' behavior change depends on their knowledge of others' behavior in their cultural circles. We moreover define a logical system suitable to reason about epistemic cultural circle based threshold models and their dynamics. To add an epistemic dimension to cultural circle based threshold models, we add for each agent a subjective epistemic indistinguishable relation like the way in [2, 3].

Definition 10 (Epistemic Cultural Circle Based Threshold Model EM). *Let a tuple $M = (W, A, B, S, N, \{C_a\}_{a\in A}, \theta, \{\sim_a\}_{a\in A})$ be an epistemic cultural circle based threshold model EM, where*

1. *W is a finite nonempty set of possible worlds.*
2. *A is a set of agent.*
3. *$\{\sim_a\} \in W \times W$ is an epistemic equivalence relation for each $a \in A$.*
4. *$S : W \longrightarrow (A \longrightarrow \mathcal{P}(A))$ assigns a social cicle Swa to each $a \in A$ in each $w \in W$ such that:*
   - *(Irreflexivity)$a \notin Swa$,*
   - *(Symmetry)$b \in Swa \Leftrightarrow a \in Swb$,*
   - *(Seriality)$Swa \neq \emptyset$.*
5. *$B : W \longrightarrow \mathcal{P}(A)$ assigns to each $w \in W$ a set $B(w)$ of agents who have adopted.*
6. *$N : A \longrightarrow \mathcal{N}$ assigns a nature number to each $a \in A$ representing the number of cultural circles her has on her social circle and $I_a = \{i|1 \leq i \leq Na\}$.*
7. *$C_a : I_a \longrightarrow \mathcal{P}(Swa)$ assigns cultural circles $\{C^i_{wa}\}_{i\in I_a}$ for each $a \in A$ in every $w \in W$ such that for each $i \in I_a$, $C^i_{wa} = C^i_{va}$ iff $Swa = Sva$.*
8. *$\theta : C_a \longrightarrow [0, 1]$ assigns a cultural circle threshold for each $C^i_a \in C_a$ of $a(\in A)$ where $C_a = \{C^i_a|i \in I_a\}$, $C^i_a = \{C^i_{wa}|w \in W\}$ and for all $i \in I_a$, $\theta_{C^i_a}$ is the same at any $w \in W$.*

In EM each possible world represent a social network with cultural circles (Definition1) and the epistemic relation is an equivalence relation. Correspondingly, for each $a \in A$ in every $w \in W$, $F_{wa} = \{X \mid \forall C^i_{wa} \in C_{Wa}, X \subseteq C^i_{wa}, |X| \geq \theta_{C^i_{wa}}|C^i_{wa}|\}$, $Ywa = \{Y \mid Y \in \mathcal{P}(Swa), X \in F_{wa}, X \subseteq Y\}$. Thus $F_{wa} = F_{va}$ and $Ywa = Yva$ if and only if $Swa = Sva$.

Definition 11 (Knowledge based Update). *Let an epistemic cultural circle based threshold model be $M = (W, A, B, S, N, \{C_a\}_{a\in A}, \theta, \{\sim_a\}_{a\in A})$, the faiths update of M generates model $M_1 = (W, A, B_1, S, N, \{C_a\}_{a\in A}, \theta, \{\sim^1_a\}_{a\in A})$, where for each $a \in A$ and all $w, v \in W$, put:*

1. *$B_1(w) = B(w) \cup \{a \in A : \forall v \sim_a w, Sva(B(v)) \in Yva\}$*
2. *$w \sim^1_a v \Leftrightarrow i)w \sim_a v$ and $ii)b \in Swa : b \in B_1(w) \Leftrightarrow b \in B_1(v)$.*

Definition 12 (Dynamic Epistemic Language KL). *Formulas in KL are given recursively as following:*

$\phi := \beta a \mid Sab \mid C^i_a ba \mid \neg\phi \mid \phi \wedge \phi \mid K_a\phi \mid [adopt]\phi$

This definition allows us to build formulas $[adopt]\phi$ ("after a round of knowledge based faith update, ϕ is the case"). The semantic interpretation of this modality refers to the knowledge based faith update model as follows:

Definition 13 (Semantic of KL). *Given an epistemic circle based threshold model $M = (W, A, B, S, N, \{C_a\}_{a\in A}, \theta, \{\sim_a\}_{a\in A})$ and the formulas in KL:*

$M, w \models \beta a$ iff $a \in B(w)$
$M, w \models Sba$ iff $b \in Swa$
$M, w \models C_a^i ba$ iff $\exists i \in I_a$ such that $b \in C_{wa}^i$
$M, w \models \neg\phi$ iff $M, w \not\models \phi$
$M, w \models \phi \wedge \psi$ iff $M, w \models \phi$ and $M, w \models \psi$
$M, w \models K_a\phi$ iff $\forall v \in W$ such that $w \sim_a v, M, v \models \phi$
$M, w \models [adopt]\phi$ iff $M_1, w \models \phi$, where M_1 is the update of M.

Note first that, the update of neighbors is the only relevant factor for knowledge based update of an agent. The unfolding of$B^* = S\beta^+a$ makes the idea more explicit: agents who will adopt do adopt after the update.

$$(B^* = S\beta^+a) := \bigwedge_{b\in B^*} (Sba \wedge [adopt]\beta b) \wedge \bigwedge_{b\notin B^*} (Sba \rightarrow [adopt]\neg\beta b) \quad (3)$$

Then the following formula captures that an agent knows thatϕ will be the case after the update if, and only if, she knows that, if those very agents who actually are going to adopt do adopt, then ϕ will hold after the update.

$$[adopt]K_a\phi \leftrightarrow \bigvee_{B^*\subseteq A} (B^* = S\beta^+a \wedge K_a(B^* = S\beta^+a \rightarrow [adopt]\phi)) \quad (4)$$

The axiom system of the logic for cultural circle based threshold models and their dynamics is presented in the following Table 1 together with the axioms in S5.

Table 1. Axioms system for KL, $(a, b \in A)$.

Axioms
$(Sba \wedge \beta b) \rightarrow K_a\beta b$
$Sba \rightarrow K_aSba$
$[adopt]Sba \leftrightarrow Sba$
$[adopt]\neg\phi \leftrightarrow \neg[adopt]\phi$
$[adopt](\varphi \wedge \psi) \leftrightarrow [adopt]\varphi \wedge [adopt]\psi$
$[adopt]\beta a \leftrightarrow \beta a \vee K_a\beta_{b\in G}$
$[adopt]K_a\phi \leftrightarrow \bigvee_{B^*\subseteq A}(B^* = S\beta^+a \wedge K_a(B^* = S\beta^+a \rightarrow [adopt]\phi))$
Inference Rules
From φ and $\varphi \rightarrow \psi$, infer ψ
From φ, infer $K_a\varphi$ for any $a \in A$
From φ, infer $[adopt]\varphi$

Definition 14 (Epistemic Logic of Cultural Circle Influence). *The logic L_{CCI} is comprised of the axioms and rules of propositional logic and S5 and the axioms and rules of Table 1, for a given threshold fuction θ, where $a \in A$.*

Definition 15 **(C_{EM}).** *For a given threshold function θ_{C_a}, where $a \in A$. the class of EM in Definition 10 is C_{EM}.*

The logic L_{CCI} is sound and complete with respect to the corresponding class of models C_{EM}.

5 Conclusions

In this article we have developed logical frameworks for the diffusion dynamics of the behavior of agents in social networks and models for the diffusion dynamics on social network with cultural circles. At start our cultural circle based threshold models did only focus on the adopting behavior of agents while in the following sections we have equipped agents with epistemic power. In the following paragraphs, we summarize our findings.

The dynamic setting of threshold models may be described sufficiently using a dynamic propositional logic with dynamic operator that indexed by actions and proposition symbols that are indexed by agents. On finite networks, threshold ratios may be encoded together with other important structural notions, such as upward closed set of influence in cultural circle and the smallest upward closed set of influence on social network. We proposed cultural circle based interpretation of faith diffusion models on social networks. As the dynamics of faith update is deterministic and state dependent, these can be described using a dynamic modality reducible to the static language. In epistemic circle based threshold models, to act as under cultural circle based threshold model dynamics, knowledge of neighbors' behavior is a necessary requirement. If this information is not available, the diffusion speed decreases or even stop in the limit case where no information is available. Thus,to give an epistemic interpretation for circle based faith updates, it is necessary to suppose that their dynamics embodies an implicit epistemic assumption that exactly the direct neighbors and their behaviors are known by agents.

It has been shown in the preliminaries that this model can simulate social behaviors inside cultural circles that the existing threshold model cannot. It is also easy to see that any threshold model can be represented as a cultural circle based threshold model in which cultural circle of an agent is consisted by all its neighbors. The same argument shows that the group conformity model also can be simulated as a special case by cultural circle based threshold model proposed in this article.

References

1. Armstrong, W.W.: Dependency structures of data base relationships. In: Information Processing 74 (Proc. IFIP Congress, Stockholm, 1974), pp. 580–583. North-Holland, Amsterdam (1974)
2. Baltag, A., Christoff, Z., Rendsvig, R.K., Smets, S.: Dynamic epistemic logics of diffusion and prediction in social networks. Stud Logica **107**(3), 489-531 (2019). https://doi.org/10.1007/s11225-018-9804-x

3. Christoff,Z., Rendsvig,R.K.: Dynamic logics for threshold models and their epistemic extension. In Epistemic logic for individual, social, and interactive epistemology workshop(2014).
4. van Ditmarsch H., van der Hoek W., Kooi B., Dynamic Epistemic Logic, Springer, Dordrecht (2008). https://link.springer.com/book/10.1007
5. Grandi, U., Lorini, E., Perrussel, L.: Propositional opinion diffusion. In: Proceedings of the 2015 International Conference on Autonomous Agents and Multiagent Systems, pp. 989–997. International Foundation for Autonomous Agents and Multiagent Systems (2015)
6. Granovetter, M.: Threshold Models of Collective Behavior. American Journal of Sociology **83**(6), 1420–1443 (1978)
7. Liu, F., Seligman, J., Girard, P.: Logical dynamics of belief change in the community. Synthese **191**(11), 2403–2431 (2014).
8. Morrison, C., Naumov, P.: Group Conformity in Social Networks, Journal of Logic Language and Information(2019) https://doi.org/10.1007/s10849-019-09303-5
9. Naumov, P., Tao, J.: Marketing impact on diffusion in social networks. Journal of Applied Logic **20**, 49–74 (2017) https://doi.org/10.1016/j.jal.2016.11.034
10. Seligman, J., Liu, F., Girard, P.: Logic in the community. In: Banerjee, M., Seth, A. (eds.) ICLA 2011. LNCS (LNAI), vol. 6521, pp. 178–188. Springer, Heidelberg (2011). https://doi.org/10.1007/978-3-642-18026-2-15
11. Seligman, J., Liu, F., Girard, P.: Facebook and the epistemic logic of friendship. In: Schipper, B.C. (ed.) TARK 2013 Proceedings of the 14th Conference on Theoretical Aspects of Rationality and Knowledge, pp. 229–238, Chennai(2013)
12. Xiong, Z., Guo, M.: A Dynamic Hybrid Logic for Followership. In: Blackburn, P.,Lorini, E., Guo, M.(Eds.): LORI 2019, LNCS 11813, pp. 425–439. (2019). https://doi.org/10.1007/978-3-662-60292-8-31

Monotonic Opaqueness in Deontic Contexts *

Jialiang Yan[1] and Fenrong Liu[2]

[1] Tsinghua University, Beijing, China
yan-jl19@mails.tsinghua.edu.cn
[2] Tsinghua University, Beijing, China
fenrong@tsinghua.edu.cn

Abstract. The paper begins with a quick review of monotonic inferences in natural language. We illustrate that the inferences sometimes fail with sentences that contain both quantifiers and deontic modalities. A distinction between a narrow and wide scope of readings is made. A first-order deontic event model is proposed to study those sentences. Inspired by the philosophy of Allan Gibbard, we set a normative system as a component of our model to interpret the modalities. This leads to a general result that enables us to explain the failed monotonic inferences.

Keywords: Monotonicity · Deontic modality · Normative judgement.

1 Introduction

Monotonicity is one of the most fundamental properties that many valid inferences depend on. Both linguists and logicians have studied reasoning with monotone quantifiers, see, e.g., [5, 8, 11].

To set the scenes, we first introduce a few basic definitions. According to [5], an *upward monotonic quantifier* of type $< t_1, ..., t_m >$ in ith argument is defined w.r.t a model M as:

> If $Q_M[A_1, ..., A_i..., A_m]$ and $A_i \subseteq A'_i \subseteq M^{n_i}$; then $Q_M[A_1, ..., A'_i, ..., A_m]$, where $1 \leq i \leq k$.

A *downward monotonic quantifier* of type $< t_1, ..., t_m >$ in ith argument is defined as:

> If $Q_M[A_1, ..., A_i, ..., A_m]$ and $A'_i \subseteq A_i \subseteq M^{n_i}$; then $Q_M[A_1, ..., A'_i..., A_m]$, where $1 \leq i \leq k$.

To understand the above definitions, we can consider the following examples including (generalized) quantifiers of type $< 1, 1 >$:

(1) a. Everyone is running. So every boy is running.
b. Everyone is running fast. So everyone is running.

* This chapter is in its final form and it is not submitted to publication anywhere else.

B. Liao and Y. N. Wáng (eds.), *Context, Conflict and Reasoning*,
Logic in Asia: Studia Logica Library,
https://doi.org/10.1007/978-981-15-7134-3_7

(2) a. Some dogs are black. So some animals are black.
b. Some dogs are black. So some animals are colored.
(3) a. None of my family members smokes. So none of my parents smokes.
b. None of my family members smokes. So none of my family members smokes on the sly.
(4) a. Not all girls have long hair. So not all people have long hair.
b. Not all girls have long hair. So not all girls have long curly blonde hair.
(5) Most adults are well educated. So most adults are educated.

In examples (1)-(4), we can reason monotonically on both noun phrases and verb phrases. In other words, the quantifiers in these examples have double monotonicity: *every* is $\downarrow$ $[Mon]$ $\uparrow$[3]; *some* is $\uparrow$ $[Mon]$ $\uparrow$; *no* is $\downarrow$ $[Mon]$ $\downarrow$; *not all* is $\uparrow$ $[Mon]$ $\downarrow$. These four quantifiers correspond to the traditional Aristotle's square of opposition that we are familiar with:

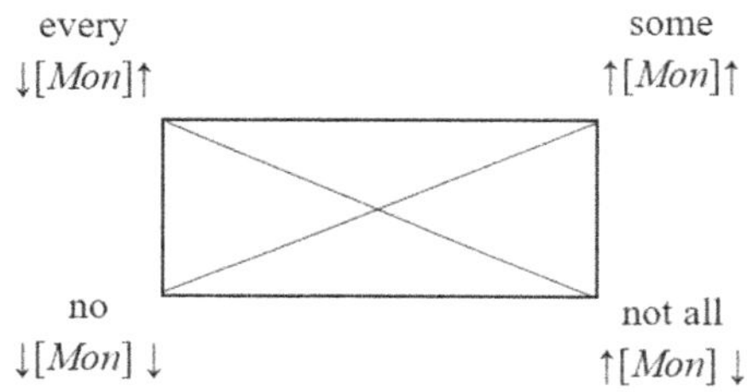

Fig. 1. Square of opposition about monotonicity.

More research on the square regarding monotonicity can be found in [12]. When we move to generalized quantifiers, things get a bit complex. Some generalized quantifiers are monotonic, e.g., *most* in the sentence (5), while some are not, e.g., *even* and *exactly two*.

In this paper, our focus will be those sentences that have both quantifiers and modalities, typically, deontic modalities. Our research question is: would the used-to-be-valid inferences that respect to the monotonicity principles of quantifiers still be valid in these modal contexts? Perhaps, a good answer we can give is 'It depends'. Consider the following four pairs of examples:

(6) a. Some murderers were arrested by Lisi. So some people were arrested by Lisi.
b. * Some murderers ought to be arrested by Lisi. So some people ought to be arrested by Lisi.

[3] The left arrow represents the monotonic inferences of noun phrases in the sentences; and the right arrow represents the monotonic inferences of verb phrases. Besides, the upward arrow denotes upward monotonicity, while the downward arrow denotes downward monotonicity.

(7) a. Some doctors benignly deceive their patients. So some doctors deceive their patients.
b. * Some doctors ought to benignly deceive their patients. So some doctors ought to deceive their patients.
(8) a. Most people follow Lei Feng[4]. So most people follow a good example.
b. Most people ought to follow Lei Feng. So most people ought to follow a good example.
(9) a. No one steals. So no boy steals.
b. No one ought to steal. So no boy ought to steal.

Obviously, (6)a–(9)a are valid according to monotonicity. In contrast, among those b-sentences, it is not clear. (8)b and (9)b seem to be valid, while (6)b and (7)b are problematic since it is debatable whether 'arrest some people' and 'deceive patients' are obligations for the agents. In (6)b and (7)b, monotonic inferences are no longer valid. Namely, principles of monotonicity do not apply. We will call such a context *monotonically opaque*. A context is monotonically opaque typically when quantifiers and modalities appear together in a sentence.

The rest of our paper is organized as follows. In Section 2, to analyze monotonically opaque contexts, we make a distinction between the wide and narrow scope of modalities. Besides, we introduce Gibbard's philosophy to discuss how to evaluate normative judgements. In Section 3, we propose a first-order deontic event model, to interpret the earlier sentences. We end up with a general result that can be adopted to explain why the inferences in monotonically opaque context fail.

2 Monotonically Opaque Contexts

2.1 Wide Scope vs. Narrow Scope

In a first-order modal sentence, the issue of scope is typical. If quantifiers lie inside the scope of a modal operator, we say that the operator has a wide scope; Otherwise, it has a narrow scope. More concretely, we can distinguish two readings of a sentence w.r.t. its scope. For example,

(10) Lisi ought to arrest some people.
a. What Lisi ought to do is to arrest some people.
b. There are some people whom Lisi ought to arrest.

(10)a is a wide scope reading of (10), and (10)b is a narrow scope reading. The individuals in (10)a have the property of being arrested by Lisi. But can we claim that arresting these individuals is what Lisi ought to do? Maybe not. Because if Lisi arrested someone who is an ordinary man, but Lisi has mistaken him as a criminal, then the ordinary man does have the property of being arrested by Lisi while he ought not to be arrested. Thus the sentence (10) seems to be

[4] Lei Feng was a disciplined, selfless and devoted PLA soldier, regarded as a model citizen that all Chinese should follow.

monotonically opaque. While the individuals in (10)b are those people who have the property that ought to be arrested by Lisi. Obviously, Lisi ought to arrest some people who ought to be arrested by him. In other words, under this reading, sentence (10) gives us complete information. Hence it is sufficient for us to make a normative judgment, and there is no monotonic opaqueness here.

Therefore we can see that the monotonic opaqueness only arises under wide scope readings. In fact, we can also think of this distinction in terms of *de dicto-de re*. The wide scope corresponds to a *de dicto* reading. The narrow scope corresponds to a *de re* reading, and the sentence expresses that some individuals have the property 'ought to be arrested'. For those who are not familiar with these two notions, we quickly cite an example from [6]:

(i) It is necessary that 9 is less than 10.
(ii) The number of the planets is 9.
(iii) It is necessary that the number of the planets is less than 10.

One cannot derive (iii) from from (i) and (ii) unless assuming that the Leibniz's law holds. The reason lies that sentence (iii) is ambiguous. We are unable to determine whether its predicate is 'less than 10', or 'necessarily less than 10'. To obviate this difficulty, we must make a distinction between the following two forms of sentence (iii):

(iii-a) 'The number of the planets' satisfies the condition that it is necessary that 'x less than 10'.
(iii-b) It is necessary that 'the number of the planets' satisfies the condition that 'x less than 10'.

(iii-a) represents *de re* modality and (iii-b) represents *de dicto* modality. When sentence (iii) is of the form (iii-b), it does not follow logically from sentence (i) and (ii). [2, 7] have more discussion on this issue and beyond.

Formally, the difference between a wide scope and a narrow scope reading can be represented as follows:

Table 1. Distinction between different readings.

Distinction	Sentence	Formula
wide scope	What Lisi ought to do is arresting some people.	$O\exists xA(lx)$
narrow scope	There is someone whom Lisi ought to arrest.	$\exists xOA(lx)$

Apparently, according to the discussion above, deontic contexts are monotonically opaque under the wide scope reading. In the next section, we will adopt Gibbard's idea to analyze its underlying causes.

2.2 Different Normative Judgments for Different Events

Now let us look at sentences (6)b-(9)b under the wide scope reading.

Clearly, sentences (6)b-(9)b contain 'ought to' expressions and they express normative judgments. According to Allan Gibbard, questions about normative judgments are questions about the rationality of some types of sentiment:

> *What a person does is morally wrong if and only if it is rational for him to feel guilty for having done it, and for others to be angry at him for having done it.* [3]

For example, consider the ancient Roman emperor Nero's murder of his mother. According to Gibbard's analysis, it is morally wrong if and only if it is rational for Nero to feel guilty at having murdered his mother, and for the rest of us to feel angry at his having done so. Nero may not have felt guilty, and indeed he did not. Nevertheless, we are not looking at whether he had the emotion; what we are concerned about is whether it was rational if he did. In Gibbard's view:

> *To think something rational is to accept norms that permit it.* [3]

If we accept a normative system that bans matricide, then it is rational to have these emotions, and we can make a normative judgment that what Nero did is wrong. This way, a normative judgment expresses an agent's acceptance of norms. Thus according to the same normative system, normative judgments of the same event will not differ from a different agent's perspective. [5] That is to say, if all of us accept the norm which forbids matricide, then we would form the same normative judgment 'Nero ought not to murder his mother'.

This theory suggests that a deontic proposition 'A ought to B' can be regarded as a factual proposition when we evaluate it with regard to a normative system. Because the truth value of a deontic proposition depends on whether the event expressed by that proposition is accepted in the normative system. In Gibbard's view, a normative system can tell which event is accepted or not. For example, consider the school regulations of Tsinghua University to be a normative system in which 'students truancy' is forbidden. Therefore the deontic proposition "every student of Tsinghua ought not to truant" can be treated as a factual proposition, since it is true in the normative system of school regulations.

Now let us consider (6)b again. Suppose that Lisi is a policeman, and fighting crime is his duty. Then the statement 'Some murderers ought to be arrested by Lisi' would be factually true since the event 'Some murderers are arrested by Lisi' is morally true. Hence, it is the consequent of (10)b that went wrong. A problem arises: arresting some people who are murderers is Lisi's duty, however, when the relative clause 'who are murderers' is removed, we are no longer be able to judge whether arresting some people is what Lisi ought to do. Employing Gibbard's theory, if Lisi is arresting some people who are ordinary men, moreover, an agent accepts a normative system that forbids Lisi to do that, then that agent is rational to feel angry when she conceives such an event. In other words, the agent's anger represents her normative judgment that Lisi ought not to arrest some people who are ordinary men. Therefore, when we are in uncertain

[5] It would be interesting to consider the situations where we allow normative systems to change. We leave this for a future investigation.

situations, we do not have enough information to make an accurate normative judgment, viz., in some situations the event will be true and in others it will be false. Consequently the statement 'some people ought to be arrested by Lisi' is not deontically true since the event 'some people are arrested by Lisi' is neither morally acceptable nor morally unacceptable.

Analogously, for (8)b, to reassure the patient, the statement 'some doctors ought to benignly deceive their patients' is a true normative judgment according to physician ethics. But we cannot make a moral judgment for the event 'some doctors deceive their patients' because we do not know the purpose of their deception.

In our opinion, the reason of the failure with (6)b or (7)b is the monotonically opaque context which is due to the lack of information. In fact, if we look at monotonic inferences from a view of model change, we will see that the set of individuals is changing and hence the information conveyed by the sentences will change as well, see more discussion on the very meaning of monotonicity in [9]. However, sometimes the changed information is not sufficient for us to evaluate the sentences, because we may not know the properties of the changed individuals, just as we do not know whether the changed individuals 'people' have the property 'ought to be arrested by Lisi'.

Based on Gibbard's ideas, a deontic proposition can be treated as a declarative proposition. Furthermore, we think it is a better way to understand normative sentences, we will make it precise in the next section.

3 DEON$_E$ Model

As shown above, in order to make a normative judgment, we need more information about the event being considered. This seems to suggest that normative considerations, in this case, override monotonicity. In this section, to discuss these issues formally, we will introduce a *first-order deontic event model* (denoted by $DEON_E$), with Gibbard's ideas as its underlying philosophy. The idea of event is partly inspired by Neo-Davidsonian event semantics.

A language $\mathcal{L}$ takes the following symbols as primitive:

Individual variables:	$x, y, z...$
Event variables :	$e_0, e_1, e_2...$
Norm variables :	$n_0, n_1, n_2...$
Constants:	$a, b, c...$
Quantifiers:	$\forall$
Propositional connectives:	$\neg, \wedge$
Modal operators:	O
n-ary$(0 \leq n)$ function symbols:	$f^n, g^n, h^n...$
n-ary relation symbols:	P^n, $R^n...$

Note we have introduced two new types of variables, event variables and norm variables. Terms of the language $\mathcal{L}$ are defined as usual. And the following formation rules specify which expressions are well-formed formulas of our language:

Definition 1 (Formula).

$$\phi ::= P(t_1, \ldots, t_n) \mid \neg\phi \mid \phi \wedge \phi \mid \forall \mathbf{v}\phi \mid O\phi$$

We adopt the standard abbreviations $\phi \rightarrow \psi = \neg(\phi \wedge \neg\psi)$, $\exists x\phi = \neg\exists x\neg\phi$. $P(t_1, \ldots, t_n)$ is atomic where t_i is a term. For the quantification, $\mathbf{v}$ could represent every variable in language $\mathcal{L}$. And the modal formula $O\phi$ expresses 'it ought to be that ϕ'.

To interpret the language, we need a domain. Given the different sorts of variables in $\mathcal{L}$, our domain will be $\mathcal{D}=D \cup E \cup N$, where D, E and N are used to interpret individual, event and norm variables, respectively. Let us first define the new component of our model: a normative system.

Definition 2 (Normative System $\mathcal{N}$). *A normative system $\mathcal{N}$ is a tuple $\langle N, E, G, f\rangle$, where*

- *N a non-empty set of norms.*
- *E a non-empty set of events.*
- *G a non-empty set of functions. Each element of G is a function $g_i : E \rightarrow D$, $1 \leqslant i \leqslant n$.*
- *A function f: $N \times E \rightarrow \{1, 0\}$ assigns 1 or 0 to an event according to a norm.*

Note that for the same event e in E, there may be several g-functions which identify all the agents who are involved in the event. Gibbard's ideas are reflected in the f-function.

Definition 3 ($DEON_E$ Model). *A $DEON_E$ model $\mathcal{M}$ is defined as a quintuple $\langle \mathcal{W}, h, \mathcal{D}, \mathcal{I}, \mathcal{N}\rangle$ where $\mathcal{W}$ is a set of possible worlds, h is a function assigning to each possible world w a non-empty set $h(w)$ of worlds and $h(w)$ represents deontically most preferred worlds from w, $\mathcal{N}$ is a normative system; $\mathcal{D}=D \cup E \cup N$, which is a non-empty domain; $\mathcal{I}$ is an interpretation over $\mathcal{D}$ that associates:*

(1) If P is an n-ary relation symbol, then $\mathcal{I}(P)$ is a mapping from $\mathcal{W}$ to subsets of $\mathcal{D}^n$.
(2) If a is a constant symbol, then $\mathcal{I}(a)$ is a partial function from $\mathcal{W}$ to an element of $\mathcal{D}$.
(3) If k is an n-ary function symbol, then $\mathcal{I}(k)$: $\mathcal{W} \rightarrow (\mathcal{D}^n \rightarrow \mathcal{D})$.

In this paper we will assume that $\mathcal{D}$ is constant cross the possible worlds, so we will omit w when it is unnecessary. A valuation v on $\mathcal{M}$ is a mapping given in the following definition:

Definition 4 (Valuation).

(1) For every individual variables x, $v(x) \in D$.

(2) For every event variables e, $v(e) \in E$.
(3) For every norm variables n, $v(n) \in N$.
(4) For every n-ary function symbol k^n, $v(k^n)=\mathcal{I}(k^n)$.

Now, let us go back to the g_i-functions in a normative system and explain how they would work. Consider the event 'David loves Lily', denoted as e, we use c_1 and c_2 to denote David and Lily in our language, respectively. With our current notations, we have $g_1(v(e)) = v(c_1)$ and $g_2(v(e)) = v(c_2)$, namely, this event has two agents involved. [6]

We define the truth condition of a formula in $\mathcal{M}$, namely, ϕ is true at w in $\mathcal{M}$ with respect to a valuation v.

Definition 5 (Truth Condition).

- $\mathcal{M}, w \Vdash_v P\ (t_1,..., t_n)$ *iff* $\langle v(t_1)(w),..., v(t_n)(w) \rangle \in \mathcal{I}(P)(w)$
- $\mathcal{M}, w \Vdash_v \neg\phi$ *iff* $\mathcal{M}, w \nVdash_v \phi$
- $\mathcal{M}, w \Vdash_v \phi \wedge \psi$ *iff* $\mathcal{M}, w \Vdash_v \phi$ *and* $\mathcal{M}, w \Vdash_v \psi$
- $\mathcal{M}, w \Vdash_v \phi \rightarrow \psi$ *iff* $\mathcal{M}, w \nVdash_v \phi$ *or* $\mathcal{M}, w \Vdash_v \psi$
- $\mathcal{M}, w \Vdash_v \forall x\phi$ *iff for every x-variant* v' *of* v, $\mathcal{M}, w \Vdash_{v'} \phi$
- $\mathcal{M}, w \Vdash_v O\phi$ *iff for every* $w' \in h(w)$ *and every term* t_i *in* ϕ, *there is an* e, $g_i(v(e)) = v(t_i)(w')$ *and* $\mathcal{M}, w' \Vdash_v \phi$; *and for every* n, $f(v(n), v(e))= 1$.[7]

In the final item, every term t_i in ϕ will be ordered by the index i in same the way as g_i.

Now let us return to the sentence 'Some murderers ought to be arrested by Lisi' in (6)b, it can be formalized under the wide scope reading as:

$$\phi\text{: } O(\exists x(M(x) \wedge A(lx)))$$

where M denotes the unary predicate 'is a murderer', A is a binary predicate, and l is a constant which represents Lisi. For any valuation v on $\mathcal{M}$:

$\mathcal{M}$, $w \Vdash_v \phi$ iff for every $w' \in h(w)$ and every term t_i in $\exists x(M(x) \wedge A(lx))$, there is an e, $g_i(v(e)) = v(t_i)(w')$ and $\mathcal{M}$, $w' \Vdash_v \exists x(M(x) \wedge A(lx))$; then for every n, $f(v(n), v(e))= 1$.

Similarly, we formalize 'Some people ought to be arrested by Lisi' as:

$$\psi\text{: } O(\exists x(A(lx)))$$

[6] In reality there might be more than two agents in one event, we then have more g_i functions. Also, the ordering of the functions is not essential.

[7] Here we adopt the universal quantifier to norms, one can also think of other options. However, in a normative system there might be conflicting norms, to ensure that an event is accepted by the normative system, it seems necessary to make every norm compatible with the event.

With respect to the valuation v on $\mathcal{M}$, we see that it is not the case that for every $w' \in h(w)$ and every term t_i in $\exists x A(lx)$, there is an e', $g_i(v(e')) = v(t_i)(w')$ and $\mathcal{M},\ w' \Vdash_v\ \exists x A(lx)$; for every n, $f(v(n), v(e')) = 1$. Thus we have $\mathcal{M},\ w \nVdash_v\ \psi$. Since we are considering the wide scope of reading, we need to check whether the formula $\exists x A(lx)$ holds in every $w' \in h(w)$. It is not necessary that the valuation of x is a murderer in every w'. In other words, it is possible for x denoting someone who is a good guy. If that is the case, the corresponding event will not be accepted by the normative system, i.e., the value of function f will not be 1.

To conclude, for the same valuation v, ϕ is true but ψ is false, hence $\phi \rightarrow \psi$ is false. So (6)b is not a valid monotonic inference. A similar argument works for (7)b. 'Some doctors ought to benignly deceive their patients' is true but the consequent is not. While for (8)b or (9)b, when its antecedent is true, its consequent has to be true, so both (8)b and (9)b are valid.

Furthermore, a general result is summarized in the following:

Table 2. Conditions of monotonic opaqueness.

e_1 in the antecedent	e_2 in the consequent	Monotonic Opaqueness
$\forall n \in N\ f(n, e_1) = 1$	$\forall n \in N\ f(n, e_2) = 1$	NO
$\forall n \in N\ f(n, e_1) = 0$	$\forall n \in N\ f(n, e_2) = 0$	NO
$\forall n \in N\ f(n, e_1) = 1$	$\exists n \in N\ f(n, e_2) \neq 1$	YES
$\forall n \in N\ f(n, e_1) = 0$	$\exists n \in N\ f(n, e_2) \neq 0$	YES

It says that if one's normative judgment of the event in the antecedent *agrees* with that of the event in the consequent, there is no monotonic opaqueness and the usual monotonicity holds. The deontic attitude dominates! We can see that (6)b and (7)b are instances of case 3; (8)b is an instance of case 1; (9)b is an instance of case 2. Moreover, it is easy to find an instance of case 4.

Except for *ought to*, there are other deontic modalities, e.g., *it is forbidden* and *it is permitted.* We claim that under a wide scope reading there exists a similar monotonic opaqueness. For instance:

(11) Every driver is forbidden to drive fast. So Every driver is forbidden to drive.
(12) No parents are permitted to spank their children. So no parents are permitted to spank their children who make a mistake.

Our analysis and conclusion can be applied to these cases.

4 Conclusion

We have considered those sentences in which quantifiers and deontic modalities occur at the same time, and we have shown that monotonically opaque context arises under a wide scope reading which corresponds to *de dicto* modality reading. In these cases, inferences with principles of monotonicity are no longer valid. To explore the reasons of failure, we employed the philosophy of Gibbard, in which a normative judgment can be reduced to a factual one by some normative systems. We proposed a new model $DEON_E$, and defined the corresponding syntax. Then we gave the truth condition for the deontic operators. The basic idea is that we look at the event that corresponds to formulas. Because normative judgments are about events, and events can be judged by normative systems. Accordingly, we can get whether the event ought to take place. If the event is not accepted, it will be morally unacceptable. Besides, we have found a general result that specifies conditions for monotonic opaqueness.

In the future, we intend to study the monotonic opaqueness in epistemic contexts and compare it with the deontic ones. If possible, we would like to provide a general framework to account for these phenomena.

Acknowledgments. This research is supported by Tsinghua University Initiative Scientific Research Program (2017THZWYX08). We are very grateful for the helpful comments from Johan van Benthem, Martin Stokhof, Mingming Liu, and the three anonymous referees of the AWPL.

References

1. Aloni, M. Individual Concepts in Modal Predicate Logic. *Journal of Philosophical Logic* 34(1), 1–64, 2005
2. Fitting. M, R.L. Mendelsohn. *First-Order Modal Logic*. Springer, 1998
3. Gibbard, A. *Wise Choices, Apt Feelings*. Harvard University Press, Cambridge, MA,1990
4. Nate, C and Matthew, C. *Deontic Modality*. Oxford University Press, 2016
5. Peters, S and Westerståhl, D. *Quantifiers in Language and Logic*. Oxford University Press, 2006
6. Quine. W.V. Notes on Existence and Necessity. *The Journal of Philosophy*, 1943.
7. Smullyan, R. Modality and Description. *The Journal of Symbolic Logic* 13(1), 31–37, 1948, .
8. Van Benthem, J. Questions About Quantifiers. *Journal of Symbolic Logic* 49(2), 443–466, 1984
9. Van Benthem, J. and Liu, F. Some Old and New Logical Aspects of Monotonicity. Accepted by *the Second Tsinghua Interdisciplinary Workshop on Logic, Language, and Meaning: Monotonicity in Logic and Language*, 2020
10. Von Fintel, K. The Best We Can (Expect to) Get? Challenges to The Classic Semantics for Deontic Modals. Presented at the 2012 Central APA, Chicago, IL. Available from: http://mit.edu/fntel/fntel-2012-apa-ought.pdf. [Accessed Nov 2019]
11. Westerståhl, D. Some Results on Quantifiers. *Notre Dame Journal of Formal Logic* 25(2), 152–170, 1984
12. Westerståhl, D. Classical vs. Modern Squares of Opposition, and Beyond. In *The Square of Opposition. A General Framework for Cognition*, 195–229, 2012

Towards a Relational Treating of Language and Logical Systems*

Lingyuan Ye[0000-0002-8983-0099]

Tsinghua University, Beijing 100084, China
yely18@mails.tsinghua.edu.cn

Abstract. Generally speaking, there are two categories of semantics theory: model-theoretic approach and proof-theoretic approach. In the first part of this paper, I will briefly analyze some inadequacies related to these two approaches, and promote an alternative relational approach, which bases semantic notions on relations between expressions. A brief discussion in general for this alternative will be provided. In the second part, I will provide a solid mathematical framework to the study of logical meanings, and show its connection with the other two approaches.

Keywords: Semantics · Meaning · Relation

1 Motivation and Introduction

In the wide varieties of philosophical and logical literature, there are basically two different ways concerning the very nature of the meaning of expressions (sentences, formulas, etc.) in a language system, viz. the model-theoretic approach and the proof-theoretic approach. The model-theoretic approach, which started from Tarski's groundbreaking work [12] and still prevails among logicians, concerns the meaning to be primarily *denotational*, or *referential*. In the standard model-theoretical approach, the denotations of atomic terms are objects, those of predicate signs are sets, those of logical connectives are truth functions, and those of sentences are truth values. However, for most model-theorists, there is still a clear distinction between the denotation of a linguistic entity and its meaning. Meaning is realized differently for different people: for Frege the senses/thoughts, for others propositions expressed by an expression, assignments for constants and variables, etc.

The second approach proceeds in a different way, which can be traced back to Gentzen [6]. Proof-theoretic approach to meaning is intrinsically *inferential*[1], for it assigns proofs or deductions as an autonomous semantic role from the very onset, rather than explaining proofs as truth-preserving procedures as it does in the model-theoretic view. To some extent, this line of thoughts can be

* This chapter is in its final form and it is not submitted to publication anywhere else.

[1] A slightly different philosophical realm called *inferentialism* puts *rules of inference* at the core of understanding the meaning, see Brandom [3]. An introduction can be found in Peregrin [8].

B. Liao and Y. N. Wáng (eds.), *Context, Conflict and Reasoning*,
Logic in Asia: Studia Logica Library,
https://doi.org/10.1007/978-981-15-7134-3_8

viewed as an example of Wittgenstein's famous dictum "meaning-is-use".[2] For example, the meaning of a logical constant is the inferential role it plays in a sequence of inference, given by its introduction or elimination rules in natural deduction or corresponding rules in Gentzen's sequent calculus. As summarized in [10], "Inferentialism and the 'meaning-as-use' view of semantics is the broad philosophical framework of proof-theoretic semantics."

There is a wide variety of semantics theories, but most of them can be categorized into these two categories. A lot of debates between these two camps have happened in the literature, both of which have been criticized by different philosophers. Here, I am only intended to refer to a few of them which I think are the most important.

An underlying (possibly serious) philosophical issue for the referential approach is the likelihood to lead to misunderstandings concerning various abstract nouns. The presupposition that for every noun there exists something it denotes to may lead to elusive ontological perspectives of specific abstract nouns. Here I quote Austin[3] [1], pp. 149:

> For 'truth' itself is an abstract noun, a camel, that is of a logical construction, which cannot get past the eye even of a grammarian. We approach it cap and categories in hand: we ask ourselves whether Truth is a substance (the Truth, the Body of Knowledge), or a quality (something like the color red, inhering in truths), or a relation ('correspondence'). But philosophers should take something more nearly their own size to strain at. What needs discussing rather is the use, or certain uses, of the word 'true'.

What Austin presents is that some abstract nouns are actually invented for grammatical convenience. Accordingly, their meaning may be only explained in terms of the conceptually simpler adjectives or verbs from which they are derived. But of course, where to draw the line between "essential words", which possess designations and "auxiliary words", which are invented to fulfill pragmatic purposes, is highly provocative.

As for the proof-theoretic semantics, its account of meaning proceeds nicely if only mathematical expressions are dealt with. In the field of mathematics, the use of proof-related notions to specify the meaning seems to be more adequate than using merely truth values. However, for a large part of other sentences, its explanatory power is limited. For example, consider the simple atomic sentence that "c is a horse". This expression is just a simple *fact*, rather than a proposition that needs to be proved. In proof-theoretic semantics, there is no systematic way to *prove* a sentence like that. On the contrary, it is extremely simple to deal with such sentences in model-theoretic semantics.

As indicated in the title of this paper, the semantics theory proposed in this paper is based on *relations*. To be specific, the philosophical idea behind it is

[2] A nice discussion for the interaction between meaning and use can be found in Queiroz [9].

[3] I draw this example from a philosophical and mathematical work about the information prospect of the foundations of quantum theory [13], pp.10.

that *the meaning of an expression within a language is fundamentally involved with the relations between that expression and other expressions in the very same language.*

An important reason why it might be instructive to think this way is a simple (and I think quite significant) observation: most activities involved with understanding the meaning happens within the range of language itself. Even when model-theorists specify the denotation of a word, most of the time they express it using words and expressions, though they make distinctions between object-language and meta-language. A concrete example is how people distinguish the meaning of two expressions. Generally speaking, a typical way people use is to introduce a "witness" and somehow show it is adequate to combine with one of the expression but inadequate with the other. Here a fundamental *compatible relation* between sentences is invoked. Another example would be the extensive use of dictionaries. The way a dictionary gives the definitions of words and phrases relies on some basic relations between different language expressions, such as equivalence, contradiction, similarity, etc. What these examples show is that *we use language itself to specify the meaning of other language entities almost at all times*, which, to a large extent, lays the relational approach on more solid ground.

However, relational approach makes a radical change in the field of semantics theory, for it violates the principle of *compositionality*, in possibly the most radical way, since the meaning of an expression now is determined by its relations with *all* the other expressions.

A brief remark on compositionality is worthy here. Actually, Frege, who is usually given the credit of proposing the principle of compositionality, had himself never said about this principle; rather, in his book *The Basics of Arithmetic* [5], he stated that: "never ask for the meaning of a word in isolation".[4] This is more in line with contextuality.

In the literature, Hodges has provided a nice proposal to reconcile the apparent conflict between compositionality and contextuality, see [7]. In Hodges' framework, denotations of expressions are Fregean values, which are, initially, partially and holistically defined. It has been proved that under particular circumstances the partial semantics can be extended to the whole set of expressions of the language compositionally. The relational approach provides another aspect of treating this issue. In [7], Hodges himself assumes a kind of constituent structure of syntax, while in the relational framework, no syntactic assumptions are made for the language. Rather, when giving a relational framework, a method to *compositionally extend* the language, say adding the conjunction of two sentences into the language, is shown. Hence, giving a relational semantics, people are able to build sentences compositionally. This particular feature will be explored in the future part of this paper.

Apart from all the differences, there are links between the relational approach and the other two. First, the proof-theoretic approach towards meaning

[4] I was provided with this example of Frege in a discussion with professor Johan van Benthem.

is to some extent in line with the relational perspective about meaning: it considers exclusively the "can-be-proved" or "is-derivable-from" relation between sentences (of course, details of the proof systems are omitted in a single relation). The relation between the relational approach and the model-theoretic approach is more intricate. They seem radically different *prima facie*, but there might be deeper links between the two. To name one, I shall give an abstract model-theoretic semantic model based on relations in section 4; but I shall confine myself not to discuss more in this direction in this paper.

In the rest part of the work, a precise mathematical framework of the particular aspect of meaning corresponding to propositional logic will be given. Of course, this is the simplest to maneuver, but it will set up a nice example of the spirit of the relational approach and how to formally express its ideas.[5]

2 Basic Definition

In the second part of this work, I propose a concrete realization of the relational approach to the study of meaning. From the start, all we have is an arbitrary set of sentences Φ, and an underlying relation $\rightleftharpoons$ on Φ. The basic relation I choose to assign on Φ is the *consistent* relation. Starting from this simple relation, several logical notions will be explored. What I basically concern here is the propositional-logical aspects of the meaning of a given language, namely, how the usual results of propositional logic can be revealed in the relational approach.

For the relation $\rightleftharpoons$ is to be understood as consistency relation, we require it to be *symmetric*. $\rightleftharpoons$ is not always reflexive, though, for the possible existence of contradictions in the language. A contradiction is inconsistent with any sentences (including itself). Apart from contradictions, $\rightleftharpoons$ is reflexive. From this basic understanding of consistency relation, it follows the definition of a *general logical frame*:

Definition 1. *The pair* $(\Phi, \rightleftharpoons)$ *is called a* ***general logical frame*** *if* Φ *is the set of sentences of some language, and* $\rightleftharpoons \subseteq \Phi \times \Phi$ *is a binary relation on* Φ *which is symmetric. In addition, for any* $\phi \in \Phi$*, either* $(\psi, \phi) \notin \rightleftharpoons$ *for all* $\psi \in \Phi$*, such* ϕ *is called a* ***contradiction*** *in* Φ*, or* $(\phi, \phi) \in \rightleftharpoons$*.*

In addition, I adapt the following notions and notations:

- If $(\psi, \phi) \in \rightleftharpoons$, call that ψ and ϕ are *consistent*, and write $\psi \rightleftharpoons \phi$.
- If $(\psi, \phi) \notin \rightleftharpoons$, call that ψ and ϕ are *inconsistent*, and write $\psi \not\rightleftharpoons \phi$.

In the following part of this paper, the symbol $\perp$ will be used to denote the set of contradictions in a given general logical frame or just a particular contradiction.

[5] One may well argue that in this paper, I do not give any more account than the proof-theoretic approach would provide, and I admit this is a fair criticism. However, the purpose of this paper is to introduce the alternative view of the relational approach, and give out *one of* its formulations focused on the propositional logical aspects of meaning. To argue for the relational approach in general, certainly more detailed work should be done.

The context will make it clear for its meaning. Also, one should notice that given a general logical frame $(\Phi, \rightleftharpoons)$, it can be the case that there are no contradictions in Φ at all, which means $\bot$ is empty.

The next special kind of sentences to consider is that of *validities*. Naturally, validities should be understood as those sentences which are consistent with every sentence except for contradictions.

Definition 2. *Given a general logical frame $(\Phi, \rightleftharpoons)$, let the set $\top$ to be the set of* ***validities****, defined as for any $\phi \in \top$, either $\phi \rightleftharpoons \psi$ or $\psi \in \bot$, for all $\psi \in \Phi$.*

Definition 2 gives us the first simple example that how different aspects of meaning can be understood relationally.

A particular point worthy of pointing out is the method used in the relational approach to achieve definitions. For those who are familiar with category theory, the method used in many categorical definitions are very similar to those used here, viz. by *universal properties*. First, it's highly contextual dependent. The apparently same sentence might have different meanings (behaviors, uses, etc.) in different settings. Besides, another feature is that the entities they define are defined "up to isomorphism". In a general logical frame, two sentences being "isomorphic" means that they cannot be distinguished by any other sentences in this frame; or in other words, they are *logically equivalent*[6].

Definition 3. *Given a general logical frame $(\Phi, \rightleftharpoons)$, say two sentenses ψ and ϕ are* ***logically equivalent****, denoted as $\psi \approx \phi$, if $\psi \rightleftharpoons \chi$ iff $\phi \rightleftharpoons \chi$, for all $\chi \in \Phi$.*

Definition 3 expresses the idea that two sentences are logically equivalent if they behave exactly the same with respect to the consistency relation. It is easy to see that the logically equivalence relation $\approx$ gives out an equivalence relation on Φ, which means it is reflexive, transitive and symmetric. The following theorem shows that validities and contradictions are equivalence classes.

Theorem 1. *Given a general logical frame $(\Phi, \rightleftharpoons)$ whose sets $\top, \bot$ are not empty, for any $\psi, \phi \in \top$, $\psi \approx \phi$; the same is for $\bot$.*

Proof. (1): Suppose $\psi, \phi \in \top$. If $\chi \notin \bot$, then by definition $\psi \rightleftharpoons \chi$ and $\phi \rightleftharpoons \chi$. If $\chi \in \bot$, then $\psi \not\rightleftharpoons \chi$ and $\phi \not\rightleftharpoons \chi$. Hence, $\psi \approx \phi$. (2): Suppose $\psi, \phi \in \bot$. Since for any $\chi \in \Phi$, χ is inconsistent with ψ and ϕ, the condition is vacuously satisfied. Hence, $\psi \approx \phi$. □

Theorem 1 confirms that all the validities and contradictions can be joined into a set respectively. If a general logical frame has non-empty sets of validities and contradictions, it will be called a *normal logical frame*.

Definition 4. *A general logical frame $(\Phi, \rightleftharpoons)$ whose sets $\top$ and $\bot$ are both not empty forms a* ***normal logical frame****, denoted as $(\Phi, \rightleftharpoons, \top, \bot)$.*

[6] Equivalence relation will be treated in more detail in section 4.

Remark 1. Once we have such meta-linguistic notions of contradictions and validities, we can *expand* our language system by introducing new language elements to make the language system more "complete". For example, we can add special elements into a general logical frame Φ and make the extended frame to be a normal one. This process can be legitimately called *normalization.*

Suppose we are given a general logical frame $(\Phi, \rightleftharpoons)$ with no contradictions and validities. It can be expanded to a normal logical frame $(\Phi', \rightleftharpoons', \top, \bot)$, by just adding two elements $\top$ and $\bot$ in the original set of sentences, which means $\Phi' = \Phi \cup \{\top, \bot\}$, and the new consistency relation $\rightleftharpoons'$ is specified as follows:

- if $\phi, \psi \in \Phi$, then $\phi \rightleftharpoons' \psi$ iff $\phi \rightleftharpoons \psi$;
- for any $\phi \in \Phi$, $\top \rightleftharpoons' \phi$ and $\phi \rightleftharpoons' \top$, and $\top \rightleftharpoons' \top$;
- for any $\psi \in \Phi'$, $\bot \not\rightleftharpoons' \psi$ and $\psi \not\rightleftharpoons' \bot$.

Based on the definition above, it is straight forward to verify that the relation $\rightleftharpoons'$ is symmetric. In addition, $\top$ is consistent with every sentence in Φ' except for $\bot$, and $\bot$ is inconsistent with every sentence in Φ', including itself. Hence, it gives out a normal logical frame.

In a relational approach to meaning, processes of this general kind, which all can be viewed as "XX-ization" where "XX" corresponds to a particular meta-property of the language, are common. I will show another example in the future part of this paper. These processes reveal a sort of "dynamic" aspect of our intention (goal, requirement, use, ...) of a language and a language itself. The general feature is that the meta-linguistic notions correspond to our intentions of using the language, and what people can do is to *completize* the language according to the meta-properties. From the relational perspective, these processes in general, signify a historical point of view of the development of languages; it also explains why such notions, such as "validities" and logical connectives, such as "and", "or", etc. are ubiquitous in different languages.[7] This historical account in some sense shows that the relational approach is at a more "fundamental" or "primary" level of giving an account of the meaning of languages.

Since, as shown above, any general logical frame can be easily expanded to a normal logical frame, I will consider only normal logical frames and I will simply call them logical frames, or just frames. I will also use Φ to denote a logical frame when it involves no confusion.

3 Logical Consequence and Conjunction

In this section, I will mainly develop how the notion of logical consequence evolves from the basic consistency relation, and also what constrains should be made to a frame in order to get a "meaningful" reasoning structure. Based on these, I will also show why we have a natural requirement for the logic connective *conjunction.*

[7] Here, the word "validity" is a *representative in a particular language*, viz. in English, of the notion of it represents. The same is for "and" and "or".

Reasoning is a big part of our daily use of language. In an extremely narrow sense, logic is the study of valid reasoning. Thus, it is helpful to discuss how the relation of logical consequence can be defined by the consistency relation.

Definition 5. *Suppose Φ is a logical frame. Say ψ is the* ***logical consequence of*** *ϕ, denoted as $\phi \vdash \psi$, if $\chi \rightleftharpoons \phi$ implies $\chi \rightleftharpoons \psi$, for all $\chi \in \Phi$.*

Definition 5 corresponds to our intuition for logical consequence. If ψ is the logical consequence of ϕ, then the "information" contained in ψ should be fully contained in ϕ, which means whenever χ is consistent with ϕ, it should also be consistent with ψ. The usual properties of logical consequence relation fit with this definition, e.g. the reflexivity and transitivity of logical consequence relation (between sentences). Also, notice that if $\phi \in \bot$, then the condition is vacuously satisfied for any $\psi \in \Phi$. Hence, the law of *Ex Falso Sequitur Quodlibet* (EFSQ) holds in this characterization, and no other sentences are able to deduce every sentence in Φ. Dually, any $\phi \in \top$ is a consequence of any $\psi \in \Phi$.

The consequence relation between sets of sentences should be considered as well, for the art of reasoning is often realized in the joint work of the premises.

Definition 6. *Suppose Γ is a* finite *subset of Φ and ϕ is a sentence in Φ. Say ϕ is the logical consequence of Γ, denoted as $\Gamma \vdash \phi$, if $\psi \vdash \gamma$ for every $\gamma \in \Gamma$ implies $\psi \vdash \phi$, for all $\psi \in \Phi$. If Γ is an infinite set, then $\Gamma \vdash \phi$ iff there exists a finite subset Σ of Γ that $\Sigma \vdash \phi$.*[8]

Definition 6 basically says that the joint power of Γ is revealed by all the sentences that Γ are able to be derived from. The requirement for Γ to be finite is due to the consideration that in most of the logical systems, sentences are constructed finitely; otherwise, there could be no such sentence that can derive every sentence in an infinite set, except for contradictions, while the infinite set is "intuitively consistent". Also, notice that for any $\phi, \psi \in \Phi$, it is the case that $\phi \vdash \psi$ iff $\{\phi\} \vdash \psi$, viz. to consider a single sentence as a singleton leads to the same result. In addition, the definition for consequence relation with infinite set of premises already results in the compactness of the logic.

However, with further consideration it turns out that the ability to do reasoning properly requires further restraints on the frame. Consider a frame $(\Phi, \rightleftharpoons)$ that $\Phi = \{p, q, \neg p, \neg q, \bot\}$ (where $\bot$ stands for a contradiction) and $\rightleftharpoons \subseteq \Phi \times \Phi$ is inherited from propositional logic. The graph of consistency relation in Φ is shown by Figure 1(a) bellow. A line between ϕ and ψ reads $\phi \rightleftharpoons \psi$. According to Definition 5, we can work out consequence relations between sentences, and this is shown in Figure 1(b). An arrow from ϕ to ψ reads $\phi \vdash \psi$. The reflexive arrows in Figure 1(b) are omitted. If we apply Definition 6, it is easy to see that for the set $\{p, q\}$, the only sentence ϕ that satisfies $\phi \vdash p$ and $\phi \vdash q$ is $\bot$. Hence, it follows that every sentence in Φ follows from the set $\{p, q\}$, which suggests that the set is inconsistent. However, in the original frame Φ, it is specified that $p \rightleftharpoons q$, which means p and q are consistent. It seems to generate a contradiction here.

[8] Letters $\Gamma, \Sigma, \Pi, \cdots$ are used to denote subsets of Φ.

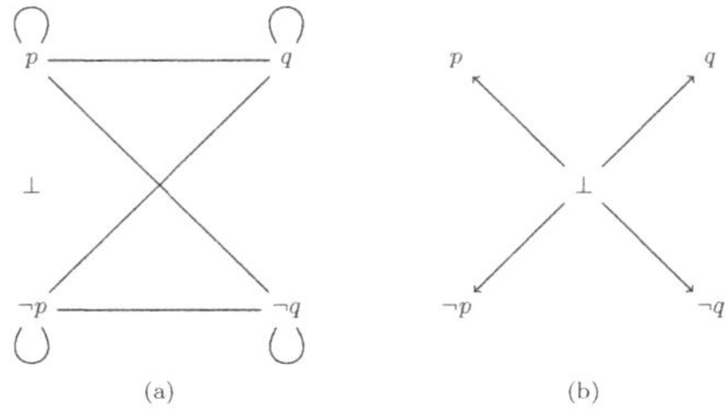

Fig. 1. Relations on Φ

The reason for the oddity of this frame is due to the fact that the relation $\rightleftharpoons \subset \Phi \times \Phi$ is "too big", in the sense that some consistency relations are not "testified". The following definition is introduced to add an adequate constraint:

Definition 7. *Suppose Φ is a logical frame. Say Φ is a* ***reasoning frame****, if for all $\phi, \psi \in \Phi$, $\phi \rightleftharpoons \psi$ iff there is a $\chi \notin \bot$ that $\chi \vdash \phi$ and $\chi \vdash \psi$.*

If we only use the binary predicate $\rightleftharpoons$, the first-order formula that characterizes this property will be expressed as the following:

$$\text{(Reasoning :)}\ \forall\phi\forall\psi(\phi \rightleftharpoons \psi \leftrightarrow \exists\chi(\chi \rightleftharpoons \chi \wedge \forall\gamma(\gamma \rightleftharpoons \chi \to (\gamma \rightleftharpoons \phi \wedge \gamma \rightleftharpoons \psi))))$$

This is a rather complicated property. It to some extent requires a "meta-consistency" relation between consistency and inference relations. For inference relation is defined using consistency relation, it ultimately sets a constraint on the consistency relation itself. Reasoning as a property never shows up in the proof of Theorem 4 below, which shows that the adequacy of our definition of inference relation is, to some extent, independent from this property.

From another perspective, we can find a natural definition for compatible relation[9] starting from inference relation as well:

Definition 8. *Given a perorder $(P, \vdash)$, the compatible relation $||$ on P is a subset of $P \times P$, where $(a, b) \in ||$ if there exists p that $\{a, b\} \not\vdash p$. Write $a||b$ for $(a, b) \in ||$.*

If we start from the consistency relation $\rightleftharpoons$ and define consequence relation $\vdash$, $\vdash$ will be a preorder, by which a compatible relation $||$ will be induced. However, generally speaking, $\rightleftharpoons$ is not the same as $||$. The frame considered above could serve as an example. From Figure 1(b) we can work out $||$ according to Definition 8. It turns out that no pair of distinct sentences are compatible with each other, which is, of course, a proper subset of $\rightleftharpoons$. This example shows that starting from inference relation is likely to "rule out" some possible consistency relations. The relation between consistency and compatible is summarized by the following theorem:

[9] Here the name "compatible relation" is used to distinguish from the consistency relation that we use through out this paper.

Theorem 2. *Suppose* $(\Phi, \rightleftharpoons)$ *is a logical frame. Let* $\vdash$ *be the consequence relation induced by* $\rightleftharpoons$*, and let* $||$ *be the compatible relation induced by* $\vdash$*. If* Φ *is a reasoning frame, then* $\rightleftharpoons$ *and* $||$ *is the same relation.*

Proof. The equivalence of $\rightleftharpoons$ and $||$ follows from the "if and only if" part of the Reasoning property.

Theorem 2 shows that possessing the Reasoning property will make consistency relation and consequence relation harmonic with each other. The analysis given above might be insufficient. There may exist deeper interaction between consistency and inference relation that worth considering both philosophically and technically, but I confine myself not to dig further. From now on, it will be assumed that all the frames are reasoning frames. Several properties related to inference relation will be investigated below.

Recall definition 3 which says that ψ and ϕ are logically equivalent iff they have the same consistency relation with respect to other formulas in the frame. Another intuition for logical equivalence relation is that ψ and ϕ are equivalent iff both of them can be derived from each other. The following theorem shows that the two conditions are the same.

Theorem 3. *Suppose* Φ *is a logical frame. For any* $\psi, \phi \in \Phi$*,* $\psi \approx \phi$ *iff* $\psi \vdash \phi$ *and* $\phi \vdash \psi$*.*

Proof. Suppose $\psi \approx \phi$. For any χ that $\chi \rightleftharpoons \psi$, since $\psi \approx \phi$, it follows that $\chi \rightleftharpoons \phi$. Hence, $\psi \vdash \phi$. Similarly, $\phi \vdash \psi$. Suppose $\psi \vdash \phi$ and $\phi \vdash \psi$. For any $\chi \rightleftharpoons \psi$, from $\psi \vdash \phi$ it follows that $\chi \rightleftharpoons \phi$. For any $\chi \rightleftharpoons \phi$, from $\phi \vdash \psi$ it follows that $\chi \rightleftharpoons \psi$. Hence, $\psi \approx \phi$. □

Both in abstract algebraic logic and other fields of logic studying, the generally accepted method of the characterization of logical consequence relation is usually attributed to Tarski (see, e.g. [4]). To show the definition above gives out a legitimate characterization of logical consequence relation, we prove the following theorem.

Theorem 4. *The logical consequence relation* $\vdash \subseteq \wp(\Phi) \times \Phi$ *(*$\wp(\Phi)$ *denotes the power set of* Φ*) has the following properties:*

(1) Reflexivity*:* $\Gamma \vdash \gamma$*, for all* $\gamma \in \Gamma$*;*
(2) Monotonicity*:* $\Gamma \vdash \phi$ *implies* $\Sigma \vdash \phi$*, for any* Σ *that* $\Gamma \subseteq \Sigma$*;*
(3) Cut*:* $\Gamma \vdash \phi$ *and* $\Sigma \vdash \gamma$ *for all* $\gamma \in \Gamma$ *implies* $\Sigma \vdash \phi$*.*

Proof. Suppose Γ, Σ are subsets of Φ.

(1): Suppose $\gamma \in \Gamma$ and Γ is finite. For any ϕ that $\phi \vdash \psi$ for any $\psi \in \Gamma$, it follows $\phi \vdash \gamma$. By definition $\Gamma \vdash \gamma$. If Γ is infinite, then for any finite subset $\Sigma \ni \gamma$, it follows that $\Sigma \vdash \gamma$. Hence, $\vdash$ is reflexive.

(2): By the definition of consequence relation, $\vdash$ is already monotone in the cases where the premises are infinite, thus only the finite cases need to be verified. Suppose $\Gamma \vdash \phi$ and $\Gamma \subseteq \Sigma$, where both Γ and Σ are finite. For any ψ that $\psi \vdash \chi$ for any $\chi \in \Sigma$, it follows that $\psi \vdash \gamma$ for any $\gamma \in \Gamma$, since $\Gamma \subseteq \Sigma$. For $\Gamma \vdash \phi$, it follows that $\psi \vdash \phi$, hence by definition $\Sigma \vdash \phi$. This proves the monotonicity.

(3): Suppose $\Gamma \vdash \phi$ and $\Sigma \vdash \gamma$ for all $\gamma \in \Gamma$.

(a). When both Γ and Σ are finite, for any ψ that $\psi \vdash \chi$ for all $\chi \in \Sigma$, by definition $\psi \vdash \gamma$ for all $\gamma \in \Gamma$. Again by definition, $\psi \vdash \phi$, hence $\Sigma \vdash \phi$.

(b). When Γ is infinite and Σ is finite, by definition there is a finite subset Γ' of Γ that $\Gamma' \vdash \phi$. For $\Gamma' \subset \Gamma$, it follows that $\Sigma \vdash \gamma'$ for all $\gamma' \in \Gamma'$. Hence by (a), $\Sigma \vdash \phi$.

(c). If Σ is infinite and Γ is finite, by definition for any $\gamma \in \Gamma$, there is a finite set $\Sigma_\gamma \subset \Sigma$ that $\Sigma_\gamma \vdash \gamma$. For Γ is finite, then the set $\cup_{\gamma \in \Gamma} \Sigma_\gamma$ is also finite. By monotonicity, $\cup_\gamma \Sigma_\gamma \vdash \gamma$ for all $\gamma \in \Gamma$. Hence, by (a), $\cup_\gamma \Sigma_\gamma \vdash \phi$. $\cup_\gamma \Sigma_\gamma$ is a finite subset of Σ, thus it follows that $\Sigma \vdash \phi$.

(d). When both Γ and Σ are infinite, then by definition there is a finite subset Γ' of Γ that $\Gamma' \vdash \phi$. For $\Gamma' \subset \Gamma$, then $\Sigma \vdash \gamma'$ for all $\gamma' \in \Gamma'$. By (c), it follows that $\Sigma \vdash \phi$.

Combining (a),(b),(c) and (d) gives the proof of (3). □

Theorem 4 shows that $\vdash$ behaves properly and it is indeed a legitimate logical consequence relation.

For another look at the definition of logical consequence relation, it can be seen that the "reasoning power" of Γ is realized by every sentence ψ that all the sentences in Γ are able to be derived from. The expectation of this definition is that, among all such ψ, there is at least one particular sentence that exactly captures the joint efforts of Γ as a set of premises[10]. In most "reasonable" logical frames, of course, this condition is satisfied. When Γ is finite, the formula that plays the same role as Γ when being treated as premises is just the *conjunction* of every sentence in Γ. Then a natural question arises, which is to ask whether there is such a single sentence that plays the exact same role as a set of sentences concerning the logical consequence relation? This concern leads to the following definition:

Definition 9. *Suppose Γ is a subset of Φ. Say a sentence is the* ***conjunction*** *of Γ, denoted as $\sqcap\Gamma$, if (1): for all $\phi \in \Phi$, $\Gamma \vdash \phi$ implies $\sqcap\Gamma \vdash \phi$; and (2): for all $\Sigma \subseteq \Phi$, $\Sigma \vdash \gamma$ for all $\gamma \in \Gamma$ implies $\Sigma \vdash \sqcap\Gamma$.*

Definition 9 is expected to capture the idea that the exact function Γ has when it is treated as a set of premises should be equivalent with the role $\sqcap\Gamma$ plays in reasoning. Condition (1) shows that $\sqcap\Gamma$ is able to derive every sentence Γ is able to derive, and condition (2) guarantees that $\sqcap\Gamma$ is not any "stronger" than Γ. The following theorem shows that the conditions can be simplified.

---

[10] Again, here the notion of "exactly captures" indicate some kind of equivalence.

Theorem 5. *Suppose Γ is a subset of Φ. Then $\sqcap\Gamma$ is the conjunction of Γ if*

(a) $\sqcap\Gamma \vdash \gamma$ for all $\gamma \in \Gamma$, and
(b) $\Gamma \vdash \sqcap\Gamma$.

Proof. (a)$\Rightarrow$(1): For any $\phi \in \Phi$ that $\Gamma \vdash \phi$, there is a finite subset Γ' of Γ that $\Gamma' \vdash \phi$. By (a), $\sqcap\Gamma \vdash \gamma'$ for all $\gamma' \in \Gamma$. Hence, $\sqcap\Gamma \vdash \phi$. (b)$\Rightarrow$(2): Suppose $\Gamma \vdash \sqcap\Gamma$. Then for any $\Sigma \subseteq \Phi$ that $\Sigma \vdash \gamma$ for all $\gamma \in \Gamma$, by cut $\Sigma \vdash \sqcap\Gamma$. □

(a) and (b) obviously follows from the definition of conjunction, since the consequence relation is reflexive. Theorem 5 shows it is sufficient to verify (a) and (b) to decide whether a sentence is the conjunction of a set of sentences or not.

Definition 9 is another instance of giving the definition using universal properties. Hence, according to definition 9, conjunction should be defined up to isomorphism (conjunction may not be uniquely defined).

Theorem 6. *Suppose Γ is a subset of Φ. If $\sqcap\Gamma_1$ and $\sqcap\Gamma_2$ both are conjunctions of Γ, then $\sqcap\Gamma_1 \approx \sqcap\Gamma_2$.*

Proof. By theorem 3, it is sufficient to verify that $\sqcap\Gamma_1 \vdash \sqcap\Gamma_2$ and $\sqcap\Gamma_2 \vdash \sqcap\Gamma_1$. It is straight foward to see both of them hold by simply applying definition. □

Recall that the motivation for looking for conjunction is to simplify a set of premises into a single sentence that preserves the properties of the original set of premises. Given a particular language system Φ and an arbitrary subset $\Gamma \subset \Phi$, $\sqcap\Gamma$ may not be unique (but all such sentences are logically equivalent), or may not exist at all, due to the property of Φ itself. In classical propositional logic, such sentences exist for every finite set $\{\phi_1, \phi_2, \cdots, \phi_n\}$. For example, $\phi_1 \wedge \phi_2 \wedge \cdots \wedge \phi_n$ is a conjunction of that set. However, conjunction for an infinite set may not exist. For example, consider the set Γ containing countably infinite many propositional letters $\Gamma = \{p_1, p_2, \cdots\}$. The only sentence that is able to derive every sentence in Γ is a contradiction $\perp$, but $\Gamma \not\vdash \perp$, for every finite subset of Γ is consistent. Hence, there does not exist a conjunction for Γ (in classical propositional logic).

If we consider other simpler logical frames, it is true that even for a finite subset, conjunction may not exist. For example, let $\Phi = \{p, q, r, s, \neg p, \neg q, \neg r, \neg s, p \wedge q \wedge r, p \wedge q \wedge s\}$, and consistency relation $\rightleftharpoons$ is inherited from propositional logic. A simple calculation will show that the subset $\{p, q\}$ has no conjunction, since both $p \wedge q \wedge r$ and $p \wedge q \wedge s$ are able to derive p and q, but neither of them is able to derive the other. The existence of such sentences for any set of sentences may be viewed as a kind of completeness of the language system itself.

Definition 10. *A logical frame Φ is said to be* ***(weakly) conjunctively complete****, if for any finite subset Γ of Φ, $\sqcap\Gamma$ exists.*

A logical frame is said to be *strongly conjunctively complete* if for any subset Γ its conjunction exists. As already been shown, conjunctively completeness is a pretty nice property (and indeed, a very basic one) for a logical system to have.

From another perspective, just like the case of validities and contradictions, given any logical frame, we may deliberately expand it by adding expressions that play such roles. For example, given any two sentences ϕ and ψ, we may add a sentence $\phi \wedge \psi$ into our set of languages, *defined* it as to be $\sqcap\{\phi, \varphi\}$. This procedure may be viewed as making a given language Φ conjunctively complete, or *conjunctive completization.*

To be more precise, given a reasoning frame $(\Phi, \rightleftharpoons)$, if ϕ and ψ are consistent, no matter their conjunction exists or not in Φ, we can add another sentence $\phi \wedge \psi$ to Φ *defined* to be their conjunction. Given a sentence π, let C_π denote the set of all sentences that are consistent with π. Let $\rightleftharpoons'$ to be the consistency relation after adding $\phi \wedge \psi$. $\rightleftharpoons \subset \rightleftharpoons'$ and for $\phi \wedge \psi$, $\rightleftharpoons'$ is given by setting

$$C'_{\phi\wedge\psi} = C_\phi \cap C_\psi \cup \{\phi \wedge \psi\}$$

where $C'_{\phi\wedge\psi}$ represents the set of all sentences that are consistent with $\phi \wedge \psi$ under the new consistency relation[11] . And for every sentence $\pi \in C_\phi \cap C_\psi$, let their consistence relation be $C'_\pi = C_\pi \cup \{\phi \wedge \psi\}$. In other words, $\rightleftharpoons'$ equals to the symmetric closure of adding $\phi \wedge \psi \rightleftharpoons' \pi$ for every $\pi \in C_\phi \cap C_\psi$ to $\rightleftharpoons$ and adding $\phi \wedge \psi \rightleftharpoons' \phi \wedge \psi$.

Theorem 7. *The sentence $\phi \wedge \psi$ given above is the conjunction of ϕ and ψ.*

Proof. Since $C'_{\phi\wedge\psi} = C_\phi \cap C_\psi \cup \{\phi \wedge \psi\}$, and $C'_\phi = C_\phi \cup \{\phi \wedge \psi\} \subset C'_{\phi\wedge\psi}, C'_\psi = C_\psi \cup \{\phi \wedge \psi\} \subset C'_{\phi\wedge\psi}$, it follows that $\phi \wedge \psi$ implies both ϕ and ψ. Given any π that $\pi \vdash \phi$ and $\pi \vdash \psi$, $C_\pi \subseteq C_\phi$ and $C_\pi \subseteq C_\psi$, hence $C'_\pi \subseteq C'_{\phi\wedge\psi}$, which follows that $\pi \vdash \phi \wedge \psi$. By definition, $\phi \wedge \psi$ is the conjunction of $\{\phi, \psi\}$. □

Theorem 7 shows the possibility of conjunctively completizing a given logical frame. As we can see here, a *compositional* linguistic entity introduced here (for the consistency relation of $\phi \wedge \psi$ is fully dependent on ϕ and ψ), while its meaning is still understood holistically. This particular point of view might shed some light on the issue between compositionality and contextuality that discussed in the introduction.

Furthermore, the techniques used in the construction of the consistency relation of $\phi \wedge \psi$ turn out to be quite useful. It is quite instructive to consider the following definition:

Definition 11. *Given a logical frame $(\Phi, \rightleftharpoons)$, there is a map $C : \Phi \to \wp(\Phi)$ defined by*

$$C(\phi) = \{\psi \in \Phi | \psi \rightleftharpoons \phi\} := C_\phi$$

called the ***embedding*** *of Φ.*

[11] It is interesting to note here that given a small frame, it is possible that there are several different ways to set up the consistency relation of $\phi \wedge \psi$. Given any π, let I_π denote the set of sentences that implies π. The set $I_\phi \cap I_\psi - \bot$ is not empty, due to the Reasoning property. Hence, another way to specify the consistency relatioin is to set $C'_{\phi\wedge\psi} = \{\phi \wedge \psi\} \cup \bigcup_{\gamma \in (I_\phi \cap I_\psi - \bot)} C_\gamma$. In fact, the consistency set given above and given here corresponds to the biggest and the smallest possible consistency relation, respectively, for $\phi \wedge \psi$. Any intermediate setting would do the job.

The set $\{C_\phi\}_{\phi\in\Phi}$ reveals the logical structure of Φ. Inference relation (between sentences) is identified with the set inclusion relation, and logical equivalence is identified with set equality. Whether the set $\{C_\phi\}_{\phi\in\Phi}$ has bounded elements or forms a sub-lattice of $\wp(\Phi)$ reveals different completeness properties of Φ (not *the completeness* in the usual sense of logic).

Having seen many "unusual" features, the following example shows that within a relatively small frame usual results in propositional logic can be generated in this framework.

Example 1. Let $\Phi = \{p, \neg p, q, \neg q, p \to q, q \to p, p \wedge q, p \wedge \neg q, \neg p \wedge q, \neg p \wedge \neg q, \bot\}$. The relation $\rightleftharpoons$ on Φ is defined by consistency relation in classical propositional logic. Starting from this consistency relation, and according to our definition of $\vdash$, similar results which all of us are already familiar with in the case of classical propositional logic will follow. For example, we may wonder if $p \wedge q \vdash p$ in this logical frame. We actually have this because it is straight forward to see whenever $p \wedge q \rightleftharpoons \phi$ for some ϕ, it must be the case that $p \rightleftharpoons \phi$. But the other way around is not true. We do not have $p \vdash p \wedge q$, because $p \rightleftharpoons \neg q$ while $p \wedge q \not\rightleftharpoons \neg q$. The relation $\vdash$ for other formulas are also easy to verify and the set is conjunctively complete. We may also wonder whether the famous Modus Ponens law holds. According to our definition, the formula $p \to q$ by itself does not imply any other formulas. However, consider the set $\{p, p \to q\}$. By definition, $\sqcap\{p, p \to q\}$ should implies both p and $p \to q$.[12] The only formula would do this job is $p \wedge q$. In addition, we have $p \wedge q \vdash q$, hence $\{p, p \to q\} \vdash q$, which shows we also have the familiar results of Modus Ponens law.

This simple example again reveals the intrinsic and essential part of the relational treatment towards meaning, namely, the meaning of a sentence within a language comes from their relationships between other languages. It is possible that distinctive expressions in a larger language cannot be distinguished within a smaller language.

4 Logical Equivalence and Meaning

In the previous section, I mainly talk about how logical consequence relation can be understood from a single consistency relation, which is in a sense in line with the proof-theoretic point of view. In this section, I will move further to the discussion of the meaning of a sentence in the scope of propositional logic, and show how the relational approach is related to the model-theoretic approach.

In order to understand the meaning of a sentence, we need, first, to understand how it is different from other sentences. In other words, two sentences have the same meaning if we cannot find a way to distinguish the two. Within the scope of propositional logic, that two sentences cannot be distinguished means they are *logically equivalent*. Logical equivalence relation has been defined in Definition 3, and it is indeed an equivalence relation on Φ. What this definition tries

[12] Notice that the meta-linguistic notion $\sqcap$ is not the same as a linguistic entity $\wedge$ within the language.

to specify is that that the (particular aspect of the) meaning of a sentence cannot be distinguished from another one means they behave exactly the same with respect to consistency relation (and Theorem 3 shows they also behave the same with respect to inference relation). Then, at least in the sense of propositional logic, they have the same meaning. However, to say two sentences have the same meaning is still one step away from giving the meaning of each sentence, but the answer is very close. What should be done is to view all the sentences that are logically equivalent as a single package, or an informational state, and use this package to represent the meaning of all these sentences. I call the remaining structure after this compression the *layer* of a language.

Mathematically, the construction of layer corresponds to the operation of constructing a quotient set $\Phi/\approx$ from Φ with respect to the equivalence relation $\approx$. As you can imagine, it is natural for the layer to inherit a consistency relation from a logical frame, and again forms a logical frame. I will show that these two logical frames are "coherent".

Definition 12. *Suppose Φ is a logical frame. The* **layer** *of Φ is the quotient set $\Phi/\approx$ of Φ under the logical equivalence relation $\approx$. Its elements are denoted as $[\phi]$, which means the equivalence class containing ϕ. The layer $\Phi/\approx$ forms a logical frame, with consistency relation $\rightleftharpoons^*$ defined as*

$$[\phi] \rightleftharpoons^* [\psi] \text{ iff } \phi \rightleftharpoons \psi$$

Under this definition, it is easy to find that the layer $\Phi/\approx$ under the relation $\rightleftharpoons^*$ forms a logical frame.

Theorem 8. *If $(\Phi, \rightleftharpoons, \top, \bot)$ is a normal logical frame, then $(\Phi/\approx, \rightleftharpoons^*, \top, \bot)$ is also a normal logical frame.*

Proof. It is a corollary from Theorem 1. □

The above properties show that the layer actually forms a logical frame, and it preserves tautology and contradiction. Starting from the new consistency relation $\rightleftharpoons^*$, a logical consequence relation $\vdash^*$ can be defined on $\Phi/\approx$. The two relations $\vdash^*$ and $\vdash$ are strongly related.

Theorem 9. *Suppose Φ is a logical frame, and $\Phi/\approx$ is the layer of Φ. Then for any $\phi, \psi \in \Phi$,*

$$[\phi] \vdash^* [\psi] \text{ iff } \phi \vdash \psi$$

Proof. According to the definition of logical consequence, $[\phi] \vdash^* [\psi]$ iff $[\phi] \rightleftharpoons^* [\varphi] \Rightarrow [\psi] \rightleftharpoons^* [\varphi]$, for all $[\varphi] \in \Phi/\approx$. According to definition of $\rightleftharpoons^*$, $[\phi] \rightleftharpoons^* [\varphi]$ iff $\phi \rightleftharpoons \varphi$ and $[\psi] \rightleftharpoons^* [\varphi]$ iff $\psi \rightleftharpoons \varphi$. It follows $[\phi] \rightleftharpoons^* [\varphi] \Rightarrow [\psi] \rightleftharpoons^* [\varphi]$, for all $[\varphi] \in \Phi/\approx$ iff $\phi \rightleftharpoons \varphi \Rightarrow \psi \rightleftharpoons \varphi$, for all $\varphi \in \Phi$. It follows $[\phi] \vdash^* [\psi]$ iff $\phi \vdash \psi$. □

Theorem 9 shows that the layer preserves the logical consequence relation between sentences. As for sets of sentences, we have a similar result.

Theorem 10. *Suppose Φ is a logical frame and Γ is a finite collection of sentences. Let $[\Gamma]$ denote the set $\{[\gamma] | \gamma \in \Gamma\}$. If $\sqcap\Gamma$ exists in Φ, then so does $\sqcap[\Gamma]$, and*

$$\sqcap[\Gamma] = [\sqcap\Gamma]$$

Proof. According to the definition, $\sqcap\Gamma \vdash \gamma$, for all $\gamma \in \Gamma$. Then according to Theorem 9, for any $[\gamma] \in [\Gamma]$, $[\sqcap\Gamma] \vdash^* [\gamma]$. For all $[\varphi] \in \Phi/\approx$, if $[\varphi] \vdash^* [\gamma]$ for all $[\gamma] \in [\Gamma]$, it follows $\varphi \vdash \gamma$ for all $\gamma \in \Gamma$. Then $\varphi \vdash \sqcap\Gamma$, which follows that $[\varphi] \vdash^* [\sqcap\Gamma]$. It follows that $[\sqcap\Gamma] = \sqcap^*[\Gamma]$. □

Corollary 1. *Suppose Φ is a logical frame. If Φ is conjunctively complete, then the layer $\Phi/\approx$ of Φ is also conjunctively complete.*

Through the above theorems and corollary, we can see the layer of a logical frame and itself has a very close connection—the important structures on a logical frame are revealed in its layer. This is what I mean by "coherent". From the perspective of proof-theoretic semantics, the meaning of a given sentence φ is fully captured by specifying what sentences (or set of sentences) can deduce φ and from φ what can we deduce, which shows what are the inferential roles φ plays. According to the previous discussion, the information related to logical consequence relation is all contained in the equivalent class φ belongs to in the layer. From a model-theoretic perspective, the layer $\Phi/\approx$ of Φ serves as a natural *semantic model* of Φ. If we view $\Phi/\approx$ as a semantic model and the elements in it as points in the model, then the satisfaction relation, or truth definition can be specified naturally.

Definition 13. *Suppose Φ is a logical frame. The* ***natural semantic model*** *of Φ is a pair $\langle \Phi/\approx, \models \rangle$, where $\Phi/\approx$ is the layer of Φ, and $\models$ is a binary relation on $\Phi/\approx \times \Phi$, defined as*

$$[\phi] \models \varphi \text{ iff } [\phi] \vdash^* [\varphi]$$

The natural semantic model given here uses elements in the layer of a language as semantic elements, with truth definition defined as φ is true in $[\phi]$ iff $[\varphi]$ is a logical consequence of $[\phi]$ in the logical frame $\Phi/\approx$.

Remark 2. This natural semantic model is not the usual model we are given with in propositional logic, while it is standard in algebraic semantics. The construction of layer is very similar to that of the Lindenbaum algebra construction, and the meta-linguistic notion of conjunction corresponds to lattice operators. However, techniques and philosophical ideas in relational semantics and algebraic logic are very different. Besides, the relational approach aims for a still higher level of abstraction than lattice theory or algebraic logic, for the relational approach does not presuppose the existence of any algebraic operators at all; that is the reason why we have such notions as conjunctively complete, etc. Hence, it is applicable to a more general class of linguistic models. It is in this sense that the relational approach is a technically more fundamental or general approach than algebraic logic.

As for one philosophical issue, to what extent the layer of a language system can be identified with its meaning is still debatable, for the natural semantic model is fully abstract. For example, in this paper what we start from is just a set and a relation on this set. It is, of course, quite possible that there are relations satisfying the various constraints given in this paper and all the constructions are applicable to them, while they have nothing to do with meaning (at least the meaning we understand). However, one thing is for sure from the results of this paper, that whatever these relations represent, at least they have the *potential* to be used as a language. And using the method specified in this paper, we can understand the sentences formed by them systematically. Thus, a safe conclusion for the relational treating of language is that it *specifies the condition of all potentially useful language systems* and *provides a systematic way of understanding the structure and the possible meaning they have based on relations.*

5 Conclusion and Future Work

Technically, what is done in this paper is starting from a set, whose elements are treated as sentences, and a relation on this set, which is understood as consistency relation. Then a series of logical notions and their properties are explored. This approach generally aims at a fundamentally more abstract way of treating different logical systems. This generality allows for further generalization of different logics. Philosophically, this framework supports the claim made in this paper that expressions can be understood within a relational treatment. Several sub-points have been drawn, including the historical aspect of completization and a possible perspective to view the contrary between compositionality and contextuality.

Besides points made above, the relational approach is also connected to other philosophical issues. For example, it might be instructive to explore the connections with the coherence picture of truth in philosophy of science, for they are similar in spirit. Technical and philosophical considerations with distributional semantics might be valuable as well.

Other technical development is also needed. For one, it seems that other connectives such as negation, disjunction, implication, etc., are also straight forward to define. Generalization to first-order logic and modal logic is also quite interesting, though possibly new relations will be introduced. In addition, more aspects of meaning should be concerned, if I am intended to reinforce my claim about the legitimacy of relational approach in general.

Acknowledgements. I would like to express my sincerest and greatest gratitude to the following professors: Fenrong Liu, Martin Stokhof and Johan van Benthem. All of them have provided me considerable help and guidance. I would also like to thank three anonymous reviewers of this paper who have also offered me many detailed and useful comments.

References

1. Austin, John L. Truth. Proceedings of the Aristotelian Society, Supp. 24, 111–129. 1950. Reprint in [2]. Page refers to this reprint.
2. Blackburn S, Simmons K. Truth Oxford University Press. 1999.
3. Brandom R. Articulating reasons: An introduction to inferentialism: Harvard University Press; 2009.
4. Cintula P, Gil-Férez J, Moraschini T, Paoli F (2019). An abstract approach to consequence relations. The Review of Symbolic Logic. 12(2):331-71.
5. Frege G. The basics of arithmetic: a logical mathematical study of the concept of number. w. Koebner; 1884th
6. Gentzen G. Untersuchungen über das logische Schließen. I. Mathematische zeitschrift. 1935;39(1):176-210. English translation 'Investigations into Logical Deduction' in [11], pp. 68–131.
7. Hodges W. 'From sentence meanings to full semantics', in Logic at the Crossroads: An Interdisciplinary View I, ed. Amitabha Gupta et al., Allied Publishers, New Delhi 2008, pp. 399-415; reprinted in Proof, Computation and Agency: Logic at the Crossroads, ed. Johan van Benthem et al., Springer, Dordrecht 2011, pp. 261-276.
8. Peregrin J. What is inferentialism? Inference, consequence, and meaning: perspectives on inferentialism. 2012:3-16.
9. De Queiroz RJ (2008). On reduction rules, meaning-as-use, and proof-theoretic semantics. Studia Logica. Nov 1;90(2):211-47.
10. Schroeder-Heister, Peter, "Proof-Theoretic Semantics", The Stanford Encyclopedia of Philosophy (Spring 2018 Edition), Edward N. Zalta (ed.), https://plato.stanford.edu/archives/spr2018/entries/proof-theoretic-semantics/.
11. Gentzen, Gerhard, and M. E. Szabo. The Collected Papers of Gerhard Gentzen. Studies in Logic and the Foundations of Mathematics. 1969.
12. Tarski A. The semantic conception of truth and the foundations of semantics. Philosophy and phenomenological research. 1944;4(3):341-76.
13. Timpson CG. Quantum information theory and the foundations of quantum mechanics: OUP Oxford; 2013.

Modal Logic and Planarity of Graphs

Izumi Takeuti[1] [0000−0001−7235−2281] and Katsuhiko Sano[2]

[1] National Institute of Advanced Industrial Science and Technology, Tsukuba, Ibaraki, 305–8560 Japan
takeuti@ni.aist.go.jp

[2] Faculty of Humanities and Human Sciences, Hokkaido University, Sapporo, Hokkaido, 060–0810 Japan
v-sano@let.hokudai.ac.jp

Abstract. When a Kripke frame is viewed as a graph, the denotation of a formula is interpreted as a set in a graph and so a modal formula can be used to describe the properties of graphs. This study particularly focuses on planarity of graphs and proposes a syntax of extended modal logic in which planarity is definable. Moreover, we provide semantically complete and decidable axiomatizations of the logics under consideration.

Keywords: Modal logic · Graph theory · Planarity · Nominals · Decidability · Universal Modality

1 Introduction

Traditionally, the meaning of a sentence is a proposition. For example, the meaning of a sentence 'The year of 2020 is a leap year.' is a proposition. On the other hand, the meaning of a formula with a hole (or one free variable) is a property. For example, the meaning of a sentence with a hole '( ) is a leap year.' is a property, so to say, leapness. In the standard model-theoretic semantics of first-order logic, a closed formula (or a sentence) is interpreted into a proposition, and its denotation is a truth value. For example, as for the sentence 'The year of 2020 is a leap year.', its denotation is 'true'. On the other hand, a formula with holes is interpreted into a property, and its denotation is a set which is its extension. For example, for the formula with a hole '( ) is a leap year.', its denotation is the set $\{x \mid x \text{ is a leap year.}\}$, which is its extension.

When we turn our eye on Kripke semantics based on graphs, the semantics interprets a formula into a set of nodes in a graph. Moreover, a modal operator refers to an operation over (the powerset of) the set of all nodes of the graph. Thus, modal logic can be used to describe properties of graphs. The aim of this study is to use modal logic for describing the property of planarity of graphs. We introduce two languages **L1** and **L2** of modal logic. The Hilbert system **Ax1** for **L1** is the same as the modal logic **B**. We show that the language **L2** can define the planarity of graphs, but the language **L1** cannot.

We proceed as follows. [3] Section 2 provides basic knowledge on graph theory. We note that our notion of graph amounts to a finite Kripke frame (W, R) such

[3] This chapter is in its final form and it is not submitted to publication anywhere else.

B. Liao and Y. N. Wáng (eds.), *Context, Conflict and Reasoning*,
Logic in Asia: Studia Logica Library,
https://doi.org/10.1007/978-981-15-7134-3_9

that R is symmetric (non-directed) and irreflexive (no loop). Section 3 introduces two langauges to talk about graphs. The first language **L1** is the syntax of propositional modal logic and the second expanded language **L2** is the expansion of **L1** with the univeral modality [6], nominals [1] and the modal operator which amounts to the relativized common knowledge [12] in our setting. Section 4 shows that planarity is undefinable in **L1** but it is definable in **L2**. Section 5 introduces two axiomatizations **Ax1** and **Ax2** for **L1** and **L2** respectively and show that they are semantically complete for the intended finite graphs hence they are decidable. Section 6 mentions future work.

Related Work

[11] studies hypergraph and graphs based on bi-intutionistic tense logic expanded with the universal modality. This paper, however, is based on classical propositional logic. [2] discusses the definability results of several properties over directed and undirected graphs. While the literature [2, Theorem 49] shows undefinability of planarity, our proof in Section 4.1 provides undefinability of planarity *within finite graphs*. So, our result is a strengthening of the result in [2]. [2] did not study the language which enables us to define planarity. While [9] emphasizes the importance of the use of hybrid logic for generalisations of graphs (i.e., coalgebras), we need for describing planarity to employ a modal operator corresponding to the relativized common knowledge [12] beyond our hybrid vocabulary, i.e., nominals and the universal modality.

2 Graph Theory

Definition 1. (Graph) *A* graph *is a pair $G = (N, E)$ where N is a non-empty finite set of* nodes *and E is a set of* edges *which are the sets of two elements of N, that is, $E \subseteq \{\{n, n'\} \mid n, n' \in N, n \neq n'\}$. We write $n \sim n'$ when $\{n, n'\} \in E$.*
For a graph $G = (N, E)$, we write $N(G)$ for N, and $E(G)$ for E.

It is noted that a graph defined above can be regarded as a Kripke frame (W, R) where W is finite and R is symmetric and irreflexive.

Definition 2. (Connectedness) *A graph $G = (N, E)$ is* connected *if, for any $n, n' \in N$, there are $n_0, n_1, ..., n_k \in N$ such that $n = n_0$, $n' = n_k$, and $n_i \sim n_{i+1}$ for each $0 \leqslant i \leqslant k - 1$.*

We write $\mathcal{P}X$ for the set of all the subsets of X.

Definition 3. (Restriction) *For a graph $G = (N, E)$ and a non-empty subset $N' \subseteq N$, the* restriction *of G by N' is the graph $(N', E \cap \mathcal{P}N')$. We write $G|_{N'}$ for the restriction of G by N'.*

A graph is said to be *planar* when there is an embedding of it into a plane without its edges' crossing. Planarity of graphs is usually defined in terms of the notion of embedding of graphs into the plane, which requires the knowledge of topology. We avoid mentioning topology to define the notion of planarity not with embedding into the plane but with Kuratowski's theorem (cf. [4, Theorem 4.4.6]). His theorem involves the notion of minor and the graphs K_5 and $K_{3,3}$. The statement of his theorem is that a graph is planar iff neither K_5 nor $K_{3,3}$ is its minor.

Definition 4. (Minor) *For graphs $G = (N, E)$ and $G' = (N', E')$, the graph G is a* minor *of G' if there is a function $f : N \to \mathcal{P}N'$ such that:*
- *$f(n) \neq \varnothing$ for each $n \in N$*
- *$G'|_{f(n)}$ is connected for each $n \in N$,*
- *$f(n) \cap f(n') = \varnothing$ for each $n, n' \in N$ where $n \neq n'$, and*
- *for each $\{n_1, n_2\} \in E$, there are $n'_1 \in f(n_1)$ and $n'_2 \in f(n_2)$ such that $n'_1 \sim' n'_2$, where $\sim'$ is defined for E'.*

Definition 5. (Complete graph, bipartite complete graph) *A* complete graph K_k *is a graph (N, E) where $N = \{n_1, n_2, ..., n_k\}$ and $E = \{\{n, n'\} \mid n, n' \in N, n \neq n'\}$. A* complete bipartite graph $K_{k,l}$ *is the graph (N, E) where*
- *$N = \{n_1, n_2, ..., n_k, n'_1, n'_2, ..., n'_l\}$, and*
- *$E = \{\{n, n'\} \mid n \in \{n_1, n_2, ..., n_k\}, n \in \{n'_1, n'_2, ..., n'_k\}\}$.*

Definition 6. (Planarity) *A graph is* planar *when neither K_5 nor $K_{3,3}$ is its minor.*

Definition 7. (Connective) *For nodes n, n' and a subset $X \subseteq N$, the set X is a* connective *from n to n' if there are $n_0, n_1, ..., n_k \in N$ where $k \geq 0$ such that the following hold:*
- *$n = n_0$, $n' = n_k$,*
- *$n_i \sim n_{i+1}$ for each $0 \leqslant i < k$, and*
- *$n_i \in X$ for each $1 \leqslant i \leqslant k$.*

We write $n \xrightarrow{X} n'$ when X is a connective from n to n'. We say that one can go from n to n' through X when $n \xrightarrow{X} n'$.

When $n \xrightarrow{X} n'$ holds, we have $(n, n') \in (\sim \cap (N \times X))^*$, where $R^* = \bigcup_{i \in \mathbb{N}} R^i$ is the reflexive transitive closure of a binary relation R. Note that it is not necessary $n' \in X$ if $n \xrightarrow{X} n'$, because it holds that $n \xrightarrow{X} n$ even if $n \notin X$.

Lemma 1. *For a graph $G = (N, E)$ and a non-empty subset $X \subseteq N$, the graph $G|_X$ is connected iff there is $n_0 \in X$ such that, for any $n \in X$, it holds that $n \xrightarrow{X} n_0$.*

3 Modal Logic on Graphs

We define two languages **L1** and **L2** of modal logic for describing structures of graphs.

Definition 8. (Formulae of L1) *The* formulae *of* **L1** *are defined by the following grammar.*

$\varphi ::= p \mid \neg\varphi \mid \varphi \wedge \varphi \mid \Box\varphi,$

where $p \in \mathsf{Var}$ *and* Var *is the countably infinite set of propositional variables. We write* $\mathit{Frml}(\mathbf{L1})$ *for the set of all formulae of* **L1**.

A formula of **L1** is interpreted into a subset of nodes with a graph $G = (N, E)$ and an evaluation $\nu : \mathsf{Var} \to \mathcal{P}N$ of propositional variables.

Definition 9. (Semantics of L1) *Under a graph* $G = (N, E)$*, for a formula* $\varphi \in \mathit{Frml}(\mathbf{L1})$ *and a function* $\nu : \mathsf{Var} \to \mathcal{P}N$*, the* interpretation *of* F *is a subset* $[\![\varphi]\!]_\nu \subseteq N$ *defined inductively as follows:*
- $[\![p]\!]_\nu = \nu(p)$ *for* $p \in \mathsf{Var}$,
- $[\![\neg\varphi]\!]_\nu = N \setminus [\![\varphi]\!]_\nu$,
- $[\![\varphi \wedge \varphi']\!]_\nu = [\![\varphi]\!]_\nu \cap [\![\varphi']\!]_\nu$,
- $[\![\Box\varphi]\!]_\nu = \{n \in N \mid \forall n' \in N.\ n \sim n' \text{ implies } n' \in [\![\varphi]\!]_\nu\}$.

Under a graph G*, we write* $n \models_\nu \varphi$ *for* $n \in [\![\varphi]\!]_\nu$.

Next, we introduce an expansion **L2** of **L1**.

Definition 10. (Formulae of L2) *The* formulae *of* **L2** *are defined by the following grammar.*

$\varphi ::= p \mid i \mid \neg\varphi \mid \varphi \wedge \varphi \mid \Box\varphi \mid \mathsf{A}\,\varphi \mid \Box^*(\varphi, \varphi),$

where $p \in \mathsf{Var}$*,* $i \in \mathsf{Nom}$*,* Var *and* Nom *are the countably infinite sets of propositional variables and nominal variables, respectively. We write* $\mathit{Frml}(\mathbf{L2})$ *for the set of all formulae of* **L2**.

Definition 11. (Semantics of L2) *Under a graph* $G = (N, E)$*, for a formula* $\varphi \in \mathit{Frml}(\mathbf{L1})$ *and a function* $\nu : \mathsf{Var} \cup \mathsf{Nom} \to \mathcal{P}N$ *such that* $\nu(i)$ *is a singleton for all* $i \in \mathsf{Nom}$*, the* interpretation *of* φ *is a subset* $[\![\varphi]\!]_\nu \subseteq N$ *defined similarly to the formulae of* **L1** *except:*
- $[\![i]\!]_\nu = \nu(i)$
- $[\![\mathsf{A}\,\varphi]\!]_\nu = \{n \in N \mid N \subseteq [\![\varphi]\!]_\nu\}$
- $[\![\Box^*(\varphi, \varphi')]\!]_\nu = \{n \in N \mid \forall n' \in N. n \overset{[\![\varphi]\!]_\nu}{\to} n' \text{ implies } n' \in [\![\varphi']\!]_\nu\}$.

Under a graph G*, we write* $n \models_\nu \varphi$ *for* $n \in [\![\varphi]\!]_\nu$.

For a formula of the form $\mathsf{A}\,\varphi$, we have the following:

$$n \models_\nu \mathsf{A}\,\varphi \text{ iff } n' \models_\nu \varphi \text{ for all } n' \in N.$$

As for a formula of the form $\Box^*(\varphi, \varphi')$, we read $\Box^*(\varphi, \varphi')$ as "for every n' such that we can go from n through $[\![\varphi]\!]_\nu$, φ' holds at n'". It is remarked that the semantic clause of $\Box^*(\varphi, \varphi')$ can be rewritten as follows:

$$n \models_\nu \Box^*(\varphi, \varphi') \text{ iff } \text{ for all } n' \in N\ ((n, n') \in (\sim \cap (N \times [\![\varphi]\!]_\nu))^* \text{ implies } n \models_\nu \varphi'),$$

which amounts mathematically to a modal operator of the relativized common knowldge [12] in this setting, where it is noted that we use the *reflexive transitive closure* instead of the transitive closure.

For $\varphi \in \mathit{Frml}(\mathbf{L1})$, the definition of $[\![\varphi]\!]$ in **L2** is exactly the same as that in **L1**. Thus, this definition of $[\![\ ]\!]$ is an extension of the former definition of $[\![\ ]\!]$. We use the following abbreviations.

Notation 1 – $\varphi \supset \psi = \neg(\varphi \wedge \neg\psi)$,
– $\varphi \equiv \psi = (\varphi \supset \psi) \wedge (\psi \supset \varphi)$,
– $\varphi \vee \psi = (\neg\varphi) \supset \psi$,
– $\Diamond\varphi = \neg\Box\neg\varphi$,
– $\mathsf{E}\,\varphi = \neg\,\mathsf{A}\,\neg\varphi$,
– $\Diamond^*(\varphi, \psi) = \neg\Box^*(\varphi, \neg\psi)$.

The order of priority is so defined that the priorities of unary operators such as $\neg$ are the highest and that of $\supset$ is the lowest. The operator $\supset$ is right associative, that is, $\varphi \supset \psi \supset \theta$ denotes $\varphi \supset (\psi \supset \theta)$. For $\Diamond^*(\varphi, \psi)$, we have the following semantic clause:

$$n \models_\nu \Diamond^*(\varphi, \varphi') \text{ iff for some } n' \in N\ ((n, n') \in (\sim \cap(N \times [\![\varphi]\!]_\nu))^* \text{ and } n \models_\nu \varphi').$$

Lemma 2. *1.* $[\![\mathsf{A}\,\varphi]\!]_\nu = N$ *iff* $[\![\varphi]\!]_\nu = N$*, and* $[\![\mathsf{A}\,\varphi]\!]_\nu = \varnothing$ *otherwise. 2.* $[\![\mathsf{E}\,\varphi]\!]_\nu = N$ *iff* $[\![\varphi]\!]_\nu \neq \varnothing$*, and* $[\![\mathsf{E}\,\varphi]\!]_\nu = \varnothing$ *otherwise.*

Definition 12. (Validity, Satisfiability) *We say that a formula φ is* valid *in a graph G and write $G \models \varphi$ when $n \models_\nu \varphi$ for each node $n \in N(G)$ and each evaluation ν under G. We say that a formula φ is* satisfiable *in a graph G and write $G \models^\exists \varphi$ when $G \not\models \neg\varphi$. We say that a formula φ is* valid *and write $\models \varphi$ when $G \models \varphi$ for each graph G.*

Definition 13. (Definability) *Let L be* **L1** *or* **L2***. A family $\mathcal{G}$ of graphs is* definable *in L when there is a set $\Gamma \subseteq \mathit{Frml}(L)$ of formulas such that, for every graph G, $G \in \mathcal{G}$ iff $G \models \varphi$ for all $\varphi \in \Gamma$. A property P over graphs is* definable *in L when the family of all the graphs which satisfy P is definable in L.*

For example, the family of the graphs which have at most two nodes is definable in **L2**, because a graph G has at most two nodes iff $G \models \mathsf{E}(i_1 \wedge i_2) \vee \mathsf{E}(i_2 \wedge i_3) \vee \mathsf{E}(i_1 \wedge i_3)$.

It is known from [5] that modal logic with nominals i and the universal modality A have the same frame definability as modal logic with the difference modality $[\neq]$ (the semantics is given as follows: $n \models [\neq]\varphi$ iff $n' \models \varphi$ for all $n' \neq n$, cf. [10]).

4 (Un-)Definability of Planarity

4.1 Spread

In this section, we define the spread of a graph and show the undefinability of planarity in **L1**. The proof of undefinability is a modification of the proof of Theorem 49 in [2] to apply to finite irrflexive graphs.

Definition 14. (Depth) *For a formula $\varphi \in Frml(\mathbf{L1})$, the* (modal) depth *of φ, which is written as $d(\varphi)$, is defined as follows:*
- $d(p) = 0$ *for* $p \in \mathsf{Var}$,
- $d(\neg\varphi) = d(\varphi)$,
- $d(\varphi \wedge \psi) = \max\{d(\varphi), d(\psi)\}$,
- $d(\Box\varphi) = d(\varphi) + 1$.

Definition 15. (Walk) *For a graph $G = (N, E)$, a node $n \in N$, a list $(n_0, n_1, ..., n_l)$ is a* walk *starting at n if the following hold:*
- $n_0 = n$,
- $n_i \in N$ *for each* $0 \leqslant i \leqslant l$, *and*
- $\{n_i, n_{i+1}\} \in E$ *for each* $0 \leqslant i \leqslant l-1$.

We use $Walk(G, n)$ to mean the set of all walks starting at n in a graph G. For a walk $w = (n_0, n_1, ..., n_l)$, the length *of w is $length(w) = l$, and the* tail *of w is $tail(w) = n_l$. For a walk $w(n_0, n_1, ..., n_l)$ and a node $n \in N$, we write $w \cdot n$ for $(n_0, n_1, ..., n_l, n)$ if it is again a walk.*

Definition 16. (Spread) *For a graph $G = (N, E)$, a node $n \in N$, and a non-negative integer d, the graph (N', E') defined below is called the* spread *of depth d at n, and we write $S(G, n, d)$ for it:*
- $N' = \{w \in Walk(G, n) \mid length(w) \leqslant d\}$,
- $E' = \{\{w, w'\} \in N' \mid w' = w \cdot tail(w')\}$.

We point out that a spread is clearly a tree, but we can also prove the following.

Lemma 3. *A spread is planar.*

Proof. We only point out that a spread is clearly a tree, and omit the details of the proof. □

Lemma 4. *Let $G = (N, E)$ be a graph, ν an evaluation over G, $n_0 \in N$ a node, and k a non-negative integer. For each formula $\varphi \in Frml(\mathbf{L1})$ such that $d(\varphi) \leqslant k$, if the length of a walk w starting at n_0 is $length(w) \leqslant k - d(\varphi)$, then the following equivalence holds: $tail(w) \models_\nu \varphi$ in G iff $w \models_{tail^{-1} \circ \nu} \varphi$ in $S(G, n_0, k)$.*

Proof. By induction on φ. □

Theorem 1. *For a graph G, a node $n \in N(G)$, an evaluation ν and a formula $\varphi \in Frml(\mathbf{L1})$, $n \models_\nu \varphi$ in G iff $n \models_{tail^{-1} \circ \nu} \varphi$ in $S(G, n, d(\varphi))$.*

Proof. By Lemma 4. □

Corollary 1. *Planarity is not definable in* **L1** *(within finite graphs).*

Proof. Let Γ be a set of formulas in $Frml(\mathbf{L1})$. Suppose that, for every graph G, G is planar iff $G \models \varphi$ for all $\varphi \in \Gamma$. Fix some non-planar graph G'. Then $G' \not\models \varphi$ for some $\varphi \in \Gamma$. Fix such $\varphi \in \Gamma$. So $G' \models^\exists \neg\varphi$. Then, by Theorem 1 and Lemma 3, there is a planar graph G'' such that $G'' \models^\exists \neg\varphi$ which is $G'' \not\models \varphi$. Therefore $G'' \not\models \varphi$ for some $\varphi \in \Gamma$. This is a contradiction with the fact that G'' is planar. □

4.2 Representation of the Minor

This section gives the representation of a minor by a formula of **L2**, and show the definability of planarity in **L2**.

Definition 17. (Representation) *Let G be a graph (N, E) where $N = \{n_1, n_2, ..., n_k\}$. Then the* representation *of G with propositional and nominal variables $p_1, p_2, ..., p_k, i_1, i_2, ..., i_k$ is the formula $R(G)$ which is the conjunction of all the following formulae:*
- $\mathsf{E}\, p_l$ *for* $1 \leqslant l \leqslant k$,
- $\mathsf{A}(p_l \supset \Diamond^*(p_l, i_l))$ *for* $1 \leqslant l \leqslant k$,
- $\mathsf{A}\, \neg(p_l \wedge p_m)$ *for* $1 \leqslant l, m \leqslant k$ *where* $l \neq m$,
- $\mathsf{E}(p_l \wedge \Diamond p_m)$ *for* $\{n_l, n_m\} \in E$.

For $\{n_i, n_j\} \in E$, the formula $R(G)$ has both $\mathsf{E}(p_l \wedge \Diamond p_m)$ and $\mathsf{E}(p_m \wedge \Diamond p_l)$, which are shown to equivalent to each other. It is noted that we include both for the convenience of the definition.

Theorem 2. *For graphs G and G', $G' \models^{\exists} R(G)$ iff G is a minor of G'.*

Proof. First, we show the left-to-right direction. Suppose $G' \models^{\exists} R(G)$, that is, $n_0 \models_\nu R(G)$ for some $n_0 \in N(G')$ and ν. Then, the following hold:
a. $[\![p_l]\!]_\nu \neq \varnothing$ for $1 \leqslant l \leqslant k$,
b. $[\![p_l]\!]_\nu \subseteq [\![\Diamond^*(p_l, i_l)]\!]_\nu$ for $1 \leqslant l \leqslant k$,
c. $[\![p_l]\!]_\nu \cap [\![p_m]\!]_\nu = \varnothing$ for $1 \leqslant l, m \leqslant k$ where $l \neq m$, and
d. $\exists n'_l \in [\![p_l]\!]_\nu \exists n'_m \in [\![p_m]\!]_\nu .\ n'_l \sim n'_m$ for $\{n_l, n_m\} \in E$.

The condition b. above says that $\forall n \in [\![p_l]\!]_\nu .\ n \overset{[\![p_l]\!]_\nu}{\to} \bar{i}$ where $\{\bar{i}_l\} = \nu(i_l)$, thus $G'|_{[\![p_l]\!]_\nu}$ is connected, by Lemma 1. The mapping $n_l \mapsto [\![p_l]\!]_\nu : N(G) \to \mathcal{P}(N(G'))$ satisfies the conditions in Definition 4. Therefore, G is a minor of G'.

Second, we prove the right-to-left direction. Suppose G is a minor of G'. Then, there is a mapping $f : N(G) \to \mathcal{P}(N(G'))$ satisfying the conditions in Definition 4. It follows that, for any $l \in \{1, 2, ..., k\}$, the set $f(n_l)$ is non-empty and connected. Thus, there is $\bar{i}_l \in f(n_l)$ such that $\forall n \in f(n_l).\ n \overset{f(n_l)}{\to} \bar{i}_l$. The evaluation ν is constructed as $\nu(i_l) = \{\bar{i}_l\}$ and $\nu(p_l) = f(n_l)$. Then, by Definition 4, the following hold:
a. $[\![p_l]\!]_\nu \neq \varnothing$ for $1 \leqslant l \leqslant k$,
b. $[\![p_l]\!]_\nu \subseteq [\![\Diamond^*(p_l, i_l)]\!]_\nu$ for $1 \leqslant l \leqslant k$,
c. $[\![p_l]\!]_\nu \cap [\![p_m]\!]_\nu = \varnothing$ for $1 \leqslant l, m \leqslant k$ where $l \neq m$, and
d. $\exists n'_l \in [\![p_l]\!]_\nu \exists n'_m \in [\![p_m]\!]_\nu .\ n'_l \sim n'_m$ for $\{n_l, n_m\} \in E$.

Therefore $n_0 \models_\nu R(G)$ for any $n_0 \in N(G')$. □

Corollary 2. *A graph G is planar iff $G \models \neg R(K_5) \wedge \neg R(K_{3,3})$. Thus, planarity is definable in* **L2**.

Proof. A graph G is planar iff neither K_5 nor $K_{3,3}$ is a minor of G, which is equivalent to $G \not\models^{\exists} R(K_5)$ and $G \not\models^{\exists} R(K_{3,3})$. This conjunction is equivalent to $G \models \neg R(K_5)$ and $G \models \neg R(K_{3,3})$ hence $G \models \neg R(K_5) \wedge \neg R(K_{3,3})$. □

5 Axiomatizations

Table 1 provides a Hilbert system **Ax1** of the language **L1**. We write $\mathbf{Ax1} \vdash \varphi$ when the system of these axioms and rules derives the formula φ.

Table 1. Hilbert System **Ax1**

(Taut)	All instances of propositional tautologies,
(MP)	From $\varphi \supset \psi$ and φ, we may infer ψ,
($\mathtt{K}_\Box$)	$\Box(\varphi \supset \psi) \supset \Box\varphi \supset \Box\psi$,
($\mathtt{B}_\Box$)	$\varphi \supset \Box\Diamond\varphi$,
($\mathtt{Nec}_\Box$)	From φ, we may infer $\Box\varphi$.

Theorem 3. *For each formula* $\varphi \in \mathit{Frml}(\mathbf{L1})$*,* $\mathbf{Ax1} \vdash \varphi$ *iff* $\models \varphi$*. Therefore,* **Ax1** *is decidable.*

Proof. Since soundness is easy, we sketch our argument for the completeness direction. We prove the contraposition and so assume that $\mathbf{Ax1} \nvdash \varphi$. It suffice for us to prove that φ is falsified in a finite irreflexive and symmetric Kripke frame, which can be regarded as a graph in this paper. Let Ξ be the all the subformulas of φ. We say that a pair (Γ, Δ) of formulas is Ξ-*complete* if $\mathbf{Ax1} \nvdash \bigwedge \Gamma \to \bigvee \Delta$ and ($\psi \in \Gamma$ or $\psi \in \Delta$) for all $\psi \in \Xi$. Define $M^{\Xi} = (W^{\Xi}, R^{\Xi}, \nu^{\Xi})$ by:

- W^{Ξ} is the set of all Ξ-complete pairs.
- $(\Gamma_1, \Delta_1) R^{\Xi} (\Gamma_2, \Delta_2)$ iff $\Box^{-1}\Gamma_1 \subseteq \Gamma_2$ and $\Box^{-1}\Gamma_2 \subseteq \Gamma_1$ where

$$\Box^{-1}\Gamma := \{\psi \mid \Box\psi \in \Gamma\}.$$

- $(\Gamma, \Delta) \in \nu(p)$ iff $p \in \Gamma$.

It is clear that R^{Ξ} is symmetric and that W^{Ξ} is finite since Ξ is finite. Now we can establish the following equivalence: For every $\psi \in \Xi$ and every $(\Gamma, \Delta) \in W^{\Xi}$, $\psi \in \Gamma$ iff $(\Gamma, \Delta) \models_{\nu^{\Xi}} \psi$. Since $\mathbf{Ax1} \nvdash \varphi$, there exists a Ξ-complete pair (Ψ, Φ) such that $\varphi \in \Phi$. By the equivalence above, we obtain $(\Phi, \Psi) \not\models_{\nu^{\Xi}} \varphi$. This implies that φ is falsified in a finite symmetric Kripke frame (W, R). But R may not be irreflexive and so we need to "bulldoze" all R-reflexive points $C := \{x \mid xRx\}$ in W by $C \times \{0, 1\}$ such that $(x, 0)$ and $(x, 1)$ can see each other (note that this construction preserves finiteness, cf. [7, p.176]). Since this construction also preserves the satisfaction, now we can conclude that φ is falsified in a finite symmetric and irreflexive Kripke frame. □

Table 2 provides a Hilbert system **Ax2** of the language **L2**. We write $\mathbf{Ax2} \vdash \varphi$ to mean that the system of these axioms and rules derives the formula φ. It is noted that the modality A satisfies all the axioms of **S5**. The underlying idea of our axiomatization of Table 2 to combine the axiomatization of modal logic with the universal modality and nominals (cf. [3, Table 5.3, p.72]) with the axioms and rules for the relative common knowledge (cf. [13, p.203], though we use a variant for the reflexive transitive closure) and the axioms for non-directedness (axiom for symmetry) and non-loopness (axiom for irreflexivity).

Table 2. Hilbert System **Ax2**

	all axioms and rules of **Ax1**
($K_{\Box^*}$)	$\Box^*(\varphi, \psi \supset \chi) \supset \Box^*(\varphi, \psi) \supset \Box^*(\varphi, \chi)$,
($\mathtt{Unfold}_{\Box^*}$)	$\Box^*(\varphi, \psi) \equiv \Box(\varphi \supset \Box^*(\varphi, \psi)) \wedge \psi$,
($\mathtt{Ind}_{\Box^*}$)	$\psi \supset \Box^*(\varphi, \psi \supset \Box(\varphi \supset \psi)) \supset \Box^*(\varphi, \psi)$,
($\mathtt{Nec}_{\Box^*}$)	From φ, we may infer $\Box^*(\psi, \varphi)$,
($\mathtt{REA}_{\Box^*}$)	From $\varphi_1 \equiv \varphi_2$, we may infer $\Box^*(\varphi_1, \psi) \equiv \Box^*(\varphi_2, \psi)$,
(K_{A})	$\mathsf{A}(\varphi \supset \psi) \supset \mathsf{A}\,\varphi \supset \mathsf{A}\,\psi$,
(T_{A})	$\mathsf{A}\,\varphi \supset \varphi$,
(B_{A})	$\varphi \supset \mathsf{A}\,\mathsf{E}\,\varphi$,
(4_{A})	$\mathsf{A}\,\varphi \supset \mathsf{A}\,\mathsf{A}\,\varphi$,
($\mathtt{Incl}_{\Box}$)	$\mathsf{A}\,\varphi \supset \Box\varphi$,
($\mathtt{Incl}_{\Box^*}$)	$\mathsf{A}\,\varphi \supset \Box^*(\psi, \varphi)$,
($\mathtt{Nec}_{\mathsf{A}}$)	From φ, we may infer $\mathsf{A}\,\varphi$,
(Nom1)	$\mathsf{E}\,i$,
(Nom2)	$\mathsf{A}(i \supset \varphi) \vee \mathsf{A}(i \supset \neg\varphi)$,
(Irr)	$i \supset \Box\neg i$.

Theorem 4. *For each formula* $\varphi \in \mathit{Frml}(\mathbf{L2})$*,* $\mathbf{Ax2} \vdash \varphi$ *iff* $\models \varphi$*. Therefore,* **Ax2** *is decidable.*

Proof. For the soundness, we just comment on (Irr). This axiom is valid, since a graph does not contain any loop. We move to the completeness and so assume that $\mathbf{Ax2} \nvdash \varphi$. It suffice for us to prove that φ is falsified in a finite irreflexive and symmetric Kripke frame, which can be regarded as a graph. Our argument below is a combination of the completness argument in [13, Section 7.8] for relativized common knowledge the completness argument in [9] for modal logic with nominals and the universal modality and our argument for Theorem 3.

Let $\mathrm{CL}(\varphi)$ (the *closure* of φ, cf. [13, Definition 7.58, p.202]) is the smallest set of formulas such that

1. $\varphi \in \mathrm{CL}(\varphi)$.
2. $\psi \in \mathrm{CL}(\varphi)$ implies $\mathrm{Sub}(\psi) \subseteq \mathrm{CL}(\varphi)$, where $\mathrm{Sub}(\psi)$ is the set of all subformulas of ψ.
3. $\Box^*(\psi, \chi) \in \mathrm{CL}(\varphi)$ implies $\Box(\psi \supset \Box^*(\psi, \chi)) \in \mathrm{CL}(\varphi)$.

We can prove that $\mathrm{CL}(\varphi)$ is finite (cf. [13, Lemma 7.59, p.203]). We put

$$\Phi := \mathrm{CL}(\varphi) \cup \{\bot\} \cup \mathrm{Sub}(\{i \supset \Box\neg i \mid i \in \mathrm{CL}(\varphi) \cap \mathsf{Nom}\})$$

where $\mathrm{Sub}(\Sigma) := \bigcup_{\sigma \in \Sigma} \mathrm{Sub}(\sigma)$. It is easy to see that Φ is still finite. Let Ξ be the set containing Φ, closed under taking $\neg$, $\supset$, $\mathsf{A}(i \supset \cdot)$ $(i \in \Phi)$ and closed under taking subformulas (cf. [9, Definition 3.9]). Then, we can prove that Ξ is finite up to logical equivalence (cf. [9, Lemma 3.10], where the logical equivalence means $\vdash \varphi \equiv \psi$). Why? Let us define $@_i\Phi := \{\mathsf{A}(i \supset \sigma) \mid \sigma \in \Phi\}$ and put $\Phi' := \mathrm{Sub}\left(\bigcup_{i \in \Phi} @_i\Phi\right)$. It is easy to see that $\Phi' \supseteq \Phi$ and that Φ' is still finite because Φ is finite. The propositional (or boolean) closure of Φ' by $\neg$ and $\supset$ is clearly finite up to logical equivalence (it becomes a finite boolean algebra). For any formula ξ in Ξ, we can find a formula σ in Φ' such that $\vdash \xi \equiv \sigma$, because the following equivalences allow us to push the occurences of $\mathsf{A}(i \supset \cdot)$ of a formula inside to hit a formula in Φ':

- **Ax2** $\vdash \mathsf{A}(i \supset \mathsf{A}(j \supset \theta)) \equiv \mathsf{A}(j \supset \theta)$,
- **Ax2** $\vdash \neg\mathsf{A}(i \supset \theta) \equiv \mathsf{A}(i \supset \neg\theta)$,
- **Ax2** $\vdash \mathsf{A}(i \supset \theta_1) \supset \mathsf{A}(i \supset \theta_2) \equiv \mathsf{A}(i \supset \theta_1 \supset \theta_2)$.

This argument ensures that Ξ is finite up to logical equivalence.

As in the proof of Theorem 3, we define the notion of Ξ-complete pair. Since we have assumed **Ax2** $\nvdash \varphi$, there exists a Ξ-complete pair (Π, Σ) such that $\varphi \in \Sigma$. We define $M^{\Xi}_{(\Pi,\Sigma)} = (W^{\Xi}, R^{\Xi}, \nu^{\Xi})$ by:

- $W^{\Xi} = \{(\Gamma, \Delta) \mid (\Gamma, \Delta) \text{ is } \Xi\text{-complete and } (\Pi, \Sigma) \sim^{\Xi}_{\mathsf{A}} (\Gamma, \Delta)\ \}$, where

$$(\Pi, \Sigma) \sim^{\Xi}_{\mathsf{A}} (\Gamma, \Delta) \text{ iff } \{\mathsf{A}\,\theta \mid \mathsf{A}\,\theta \in \Pi\} \subseteq \Gamma \text{ and } \{\mathsf{A}\,\theta \mid \mathsf{A}\,\theta \in \Gamma\} \subseteq \Pi.$$

- $(\Gamma_1, \Delta_1) R^{\Xi} (\Gamma_2, \Delta_2)$ iff $\Box^{-1}\Gamma_1 \subseteq \Gamma_2$ and $\Box^{-1}\Gamma_2 \subseteq \Gamma_1$ where

$$\Box^{-1}\Gamma := \{\psi \mid \Box\psi \in \Gamma\}.$$

- $(\Gamma, \Delta) \in \nu(p)$ iff $p \in \Gamma$.

It is remarked that ν^{Ξ} is an evaluation, i.e., $\nu^{\Xi}(i)$ is a singleton for all $i \in \Xi$. This is because $|\nu^{\Xi}(i)| \leqslant 1$ holds by axiom (`Nom2`) and $|\nu^{\Xi}(i)| \geqslant 1$ holds by axiom (`Nom1`)). Since Ξ is finite up to logical equivalence, we can assure that W^{Ξ} is finite. This can be shown as follows: It suffices to say that the number of all possible Ξ-complete pairs is finite. Let (Γ, Δ) be a Ξ-complete pair. We note that $\vdash \sigma_1 \equiv \sigma_2$ implies the equivalence $\sigma_1 \in \Gamma$ iff $\sigma_2 \in \Gamma$. Since $\Xi = \Gamma \cup \Delta$ and $\Gamma \cap \Delta = \varnothing$, the number of all possible Ξ-pairs is bounded by the number of all partitions of the quotient (finite) set of Ξ with respect to the logical equivalence.

Now we can establish the following equivalence: For every $\psi \in \Xi$ and every $(\Gamma, \Delta) \in W^{\Xi}$, $\psi \in \Gamma$ iff $(\Gamma, \Delta) \models_{\nu^{\Xi}} \psi$ in $M^{\Xi}_{(\Pi,\Sigma)}$.

We comment on the case where ψ is of the form $\Box^*(\chi, \theta)$. Then the following are equivalent (cf. [13, Lemma 7.60, p.203]):

1. $\Box^*(\chi, \theta) \in \Gamma$.
2. $\theta \in \Gamma$ and, for every $(\Gamma', \Delta') \in W^{\Xi}$ and every χ-path π from (Γ', Δ'), if $(\Gamma, \Delta) R^{\Xi} (\Gamma', \Delta')$ then π is a θ-path,

where a ψ-path is a finite sequence $(\Gamma_0, \Delta_0), \ldots, (\Gamma_n, \Delta_n)$ in W^{Ξ} such that $\psi \in \Gamma_k$ for all $0 \leqslant k \leqslant n$, $\Gamma_k R^{\Xi} \Gamma_{k+1}$ for all $0 \leqslant k < n$ and the length of the sequence $(\Gamma_0, \Delta_0), \ldots, (\Gamma_n, \Delta_n)$ is defined to be n. By the equivalence above, we obtain $(\Pi, \Sigma) \not\models_{\nu^{\Xi}} \varphi$. This implies that φ is falsified in a finite symmetric Kripke frame (W, R). But R may not be irreflexive (this may happen for $(\Gamma, \Delta) \in W^{\Xi}$ when Γ does not contain any nominals) and so we need to apply the same construction (bulldozing R^{Ξ}-reflexive points) as in the proof of Theorem 3. The construction still preserves the satisfaction relation even if we have nominals, the universal modality and $\Box^*(\cdot, \cdot)$. Thus we can conclude that φ is falsified in a finite symmetric and irreflexive Kripke frame. $\Box$

Corollary 3. *Hilbert system* **Ax2** *is a conservative extension of Hilbert system* **Ax1**, *that is, for each* $\varphi \in$ *Frml*(**L1**), **Ax1** $\vdash \varphi$ *iff* **Ax2** $\vdash \varphi$.

Proof. For each $\varphi \in$ *Frml*(**L1**), if **Ax2** $\vdash \varphi$, then $\models \varphi$ by soundness of **Ax2**, thus **Ax1** $\vdash \varphi$ by completeness of **Ax1**. $\Box$

6 Furthur Direction

The modalities $\Diamond^*(\cdot, \cdot)$ and A are definable in μ-calculus [8] in the following sense:

$\Diamond^*(\varphi, \psi) = \mu x.\psi \vee \Diamond(\varphi \wedge x)$,

$\mathsf{A}\,\varphi = \nu x.\varphi \wedge \Box x \wedge \Box' x$,

where $\Box'$ is the modality corresponding to the relation which relates disconnected components to each other. Both have only isolated bindings, that is, bound variables never occur in the scopes of other binders. Finite model property holds on μ-calculus [14]. Therefore, we have the following conjecture:

Conjecture 1. If a formula with only isolated bindings is satisfiable in μ-calculus, then it is satisfied by a planar graph.

If this conjecture holds, then planarity is not definable in the logic **L1** with modalities A and $\Diamond^*(\cdot, \cdot)$, and we can state that nominals are essential to define planarity. Under this conjecture, we have the observation such that, without the mechanism to designate unique nodes, we could avoid crossing of edges by copying some part in on side of an edge into the other side. It is a future work to consider this conjecture.

Acknoledgement

The authors are grateful for the comments by three anonymous referees. The work of the second author was partially supported by JSPS KAKENHI Grant-in-Aid for Scientific Research (C) Grant Number 19K12113, JSPS KAKENHI Grant-in-Aid for Scientific Research (B) Grant Number 17H02258, and JSPS Core-to-Core Program (A. Advanced Research Networks).

References

1. Areces, C., ten Cate, B.: Hybrid logics. In: Blackburn, P., van Benthem, J., Wolter, F. (eds.) Handbook of Modal Logic, pp. 821–868. Elsevier (2007).
2. Benevides, M.: Modal logics for finite graphs. In: Logic for Concurrency and Synchronisation, pp. 239–267. Kluwer Academic Publishers, Norwell (2003)
3. ten Cate, B.: Model theory for extended modal languages. Ph.D. thesis, University of Amsterdam, Institute for Logic, Language and Computation (2005)
4. Diestel, R.: Graph Theory. Springer, 5 edn. (2017)
5. Gargov, G., Goranko, V.: Modal logic with names. Journal of Philosophical Logic **22**, 607–636 (1993).
6. Goranko, V., Passy, S.: Using the universal modality: Gains and questions. Journal of Logic and Computation **2**(1), 5–30 (1992).
7. Hughes, G.H., Cresswell, M.J.: A New Introduction to Modal Logic. Routledge, London and New York (1996)
8. Kozen, D.: Results on the propositional μ-calculus. Theoretical Computer Science, **27**(3), 333–354 (1983)
9. Myers, R., Pattinson, D.: Hybrid logic with the difference modality for generalisations of graphs. Journal of Applied Logic **8**(4), 441–458 (December 2010).
10. de Rijke, M.: The modal logic of inequality. Journal of Symbolic Logic **57**, 56–84 (1992).
11. Sindoni, G., Sano, K., Stell, J.G.: Axiomatizing discrete spatial relations. In: Desharnais, J. et al. (ed.) Relational and Algebraic Methods in Computer Science. vol. 11194, pp. 1–18 (2018)
12. van Benthem, J., van Eijke, J., Kooi, B.: Logics of communication and change. Information and Computation **204**, 1620–1662 (2006).
13. van Ditmarsch, H., van der Hoek, W., Kooi, B.: Dynamic Epistemic Logic. Springer (2008)
14. Walukiewicz, I.: A complete deductive system for the μ-calculus. In: Proceedings of the Eighth Annual IEEE Symposium on Logic in Computer Science. pp. 136–147. IEEE Computer Science Press (1993)

Proof-theoretic Results of Common Sense Modal Predicate Calculi ⋆

Takahiro Sawasaki[1] and Katsuhiko Sano[2]

[1] Graduate School of Letters,
Hokkaido University, Sapporo, Hokkaido, Japan
taka.sawasaki562@gmail.com
[2] Faculty of Humanities and Human Sciences,
Hokkaido University, Sapporo, Hokkaido, Japan
v-sano@let.hokudai.ac.jp

Abstract. The paper presents Hilbert-style systems and sequent calculi for some weaker versions of common sense modal predicate calculus. The main results are the strong completeness results for the Hilbert-style systems and cut elimination theorems for the sequent calculi.

Keywords: common sense modal predicate logic · modal predicate logic · Hilbert-style system · sequent calculus

Introduction

Modal predicate logics have often been presented as either assuming constant domain or increasing domain. The first assumption is that whatever exists in a world exists in every world and the second assumption is that whatever exists in a world exists in every world accessible from the world. However, these assumptions are less acceptable from a philosophical viewpoint. For example, contrary to each assumption, we sometimes say that we might not have been born or that the building in front of us might be torn down and cease to exist one day. Of course, there have been a number of attempts in the literature to properly understand our talk on modality under these assumptions, but none of them seem to have been widely accepted as a solution. For comprehensive discussions on the topic, see [7, pp. 274–311], [3, pp. 81–185], [4], etc.

Common sense Modal Predicate Calculus (CMPC), which has been proposed by J. van Benthem in [1, pp. 120–121] and further developed by J. Seligman in [14,16,15], does not depend on any such assumptions. The first proponent (as far as we know), van Benthem, makes these assumptions optional, not for philosophical reasons, but "to see what proposed axioms mean in terms of frame correspondence" [1, p. 121]. On the other hand, Seligman drops these assumptions for a philosophical reason, i.e., to "take $\exists$ to mean just 'exists' while denying the Constant Domain thesis" [14, p. 8]. Instead of adopting these assumptions, both of these authors make changes to the satisfaction relation of a formula $\Box\varphi$ in such a way that we must talk about *only things in each*

⋆ This chapter is in its final form and it is not submitted to publication anywhere else.

B. Liao and Y. N. Wáng (eds.), *Context, Conflict and Reasoning*,
Logic in Asia: Studia Logica Library,
https://doi.org/10.1007/978-981-15-7134-3_10

world in which they exist. The semantics given there does not validate axiom schema K, instead it validates the following axiom schema K_{inv} provided by Seligman:

$$\mathrm{K}_{inv} \quad \Box(\varphi \supset \psi) \supset (\Box\varphi \supset \Box\psi) \qquad \textit{if all free variables in } \varphi \textit{ are also free in } \psi.$$

It should be also pointed out that A. Hazen was aware in [6] that D. Lewis' counterpart-theoretic semantics does not validate K for much the same reason. Interestingly, unlike Seligman, Hazen considers the invalidity of K to be a "serious failing" in Lewis' semantics [6, p. 326].

There still remains room for further study in CMPC. Firstly, CMPC should be more studied for its own sake. For example, as the CMPC axiomatized by Seligman is a version of **S5** with K_{inv}, some "basic" CMPCs like K_{inv}-restricted CMPC (the logic obtained from first-order logic by adding only K_{inv} and the necessitation rule) have not been examined yet. Proof theory of such logics is also still worth studying, since neither natural deductions nor sequent calculi for such logics have been proposed. The current paper thus proposes Hilbert-style systems and sequent calculi for K_{inv}-restricted CMPC and its extensions with axiom schemata T, D and analogs of D.

Secondly, frame definability in terms of the syntax of CMPC also deserves to be studied further. Van Benthem [1] studies model theory of modal predicate logic based on the syntax of CMPC, but the main syntactic focus there is the interaction between quantifiers and modality. The corresponding frame properties to familiar axioms such as T, D, etc. have not been investigated in [1]. Surprisingly, however, this is not so trivial. For example, as we will see in Proposition 9, $\Box P \supset \Diamond P$ defines the ordinary seriality, but $\Box Px \supset \Diamond Px$ does not.

The paper proceeds as follows. We first stipulate syntax for CMPCs and introduce semantics for them in Section 1. We also provide frame definability results (Proposition 9) which have not been examined yet in [1,14,16,15]. Then, we propose Hilbert-style systems for CMPCs and show the strong completeness results for them (Theorem 1) in Sections 2 and 3 respectively. We finally present sequent calculi for CMPCs and show cut elimination theorems for them (Theorem 2) in Section 4.

1 Syntax and Semantics for CMPCs

The *language* $\mathcal{L}$ of *common sense modal predicate calculi* (CMPCs) consists of a countable set $\mathsf{Var} = \{x, y, \dots\}$ of *variables*, a countable set $\mathsf{Pred} = \{P, Q, \dots\}$ of *predicate symbols* each of which has a fixed finite arity, and the set of *logical constants*: $\neg, \supset, \Box$, and $\forall$. Instead of $\mathcal{L}$, we also write $\mathcal{L}(\mathsf{Var})$ to explicitly represent the set Var of all variables in the language. The set Form of *formulas* is defined as follows:

$$\mathsf{Form} \ni \varphi ::= Px_1 \dots x_n \mid \neg\varphi \mid (\varphi \supset \varphi) \mid \forall x\varphi \mid \Box\varphi,$$

where P is a predicate symbol with arity n and $x, x_1, \dots, x_n$ are variables. The logical constants $\bot$ and $\top$ are defined as $\bot := \neg(P \supset P)$ for some fixed predicate symbol P with arity 0 and $\top := \bot \supset \bot$, the modal operator $\Diamond$ is defined as $\Diamond := \neg\Box\neg$, and the other connectives $\wedge, \vee, \exists, \supset\subset$ are defined as usual. Given a set $\Gamma \cup \{\varphi\}$ of formulas, we define the sets $\mathsf{FV}(\varphi)$ and $\mathsf{FV}(\Gamma)$ of free variables in φ and Γ, respectively as usual. We also define substitutions $z[y/x]$ and $\varphi[y/x]$ of a variable y for a variable x in a variable z and a formula φ respectively as usual, where any bound variables in φ are relabelled, if necessary, to avoid clashes. In addition, we stipulate $\mathsf{Form}(\Gamma)$ as $\{\psi \in \mathsf{Form} \mid \mathsf{FV}(\psi) \subseteq \mathsf{FV}(\Gamma)\}$.

A *frame for CMPCs* is a tuple $F = (W, \{D_w\}_{w \in W}, R)$, where W is a nonempty set whose elements are called *worlds*; each D_w is a nonempty set of *objects* and called *domain of* w; R is a binary relation on W. Note that neither of the conditions that $D_w = D_v$ for any worlds w, v (constant domain) and that wRv implies $D_w \subseteq D_v$ for any worlds w, v (increasing domain) are required. Thus, each example displayed in Figure 1 is a frame for CMPCs.

Fig. 1. Frames for CMPCs

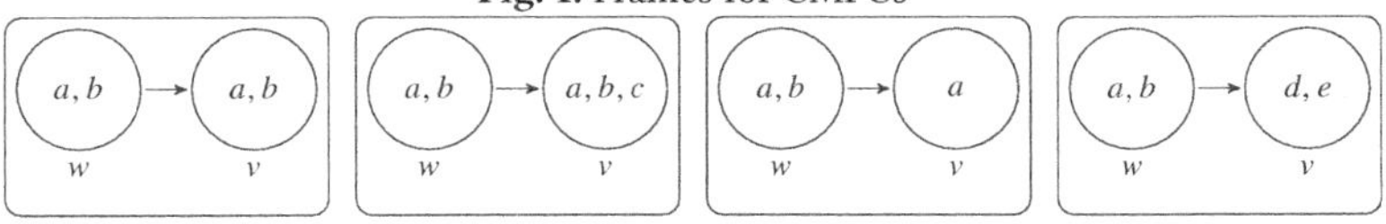

A *model for CMPCs* is a tuple $M = (F, V)$, where F is a frame for CMPCs and V is a valuation that maps each world w and each predicate symbol P with arity n to a subset $V_w(P)$ of D_w^n. An *assignment* α is a function from Var to $\bigcup_{w \in W} D_w$. The assignment $\alpha(x|d)$ stands for the same assignment as α except for assigning d to x. In addition to these notions, we follow [14, p. 15] to say that a formula φ is an α_w*-formula* if $\alpha(x) \in D_w$ for any variable $x \in \mathsf{FV}(\varphi)$. Similarly as in [14, pp. 15–16], we define the satisfaction relation and validity as follows.

Definition 1 (Satisfaction relation). *Let M be a model for CMPCs, α be an assignment, and w be a world in W. The* satisfaction relation $M, \alpha, w \models \varphi$ *between M, α, w and an α_w-formula φ is defined as follows.*

$M, \alpha, w \models Px_1 \ldots x_n$	*iff*	$(\alpha(x_1), \ldots, \alpha(x_n)) \in V_w(P)$
$M, \alpha, w \models \neg\psi$	*iff*	$M, \alpha, w \not\models \psi$
$M, \alpha, w \models \psi \supset \gamma$	*iff*	$M, \alpha, w \models \psi$ *implies* $M, \alpha, w \models \gamma$
$M, \alpha, w \models \forall x\psi$	*iff*	$M, \alpha(x|d), w \models \psi$ *for any object* $d \in D_w$
$M, \alpha, w \models \Box\psi$	*iff*	$M, \alpha, v \models \psi$ *for any world* v *such that* wRv *and* ψ *is an* α_v*-formula*

Note that only the satisfaction relation for $\Box\psi$ is peculiar of CMPCs. The intuitive meaning of the right hand side is "ψ is true at v for any world v such that w can see v and ψ is well-defined in v." Note also that $M, \alpha, w \models \Diamond\psi$ iff $M, \alpha, v \models \psi$ for some world v such that wRv and ψ is an α_v-formula. van Benthem's satisfaction relation for $\Diamond\psi$ in [1, p. 121] is the same as Seligman's in its unfolded form: $M, \alpha, w \models \Diamond\psi$ iff $M, \alpha, v \models \psi$ for some world v such that wRv and $\alpha(x) \in D_v$ for any variable $x \in \mathsf{FV}(\psi)$.

Definition 2 (Validity). *Let $\Gamma \cup \{\varphi\}$ be a set of formulas. We say that φ* is valid in a frame F for CMPCs, *denoted by $F \models \varphi$, if for any model M based on F, world w and* *assignment* α *such that* φ *is an* α_w*-formula,* $M, \alpha, w \models \varphi$*. We also say that* φ is valid in a class $\mathbb{F}$ of frames *if* $F \models \varphi$ *for any frame* F *in* $\mathbb{F}$*. We denote by* $M, \alpha, w \models \Gamma$ *that* $M, \alpha, w \models \psi$ *for all* $\psi \in \Gamma$ *and say that* φ is a consequence from Γ in $\mathbb{F}$ *if* $M, \alpha, w \models \Gamma$ *implies* $M, \alpha, w \models \varphi$ *for any model* M *based on a frame in* $\mathbb{F}$*, world* w*, assignment* α *such that* ψ *is an* α_w*-formula for all* $\psi \in \Gamma \cup \{\varphi\}$.

Lemma 1 (Substitution Lemma). *Let* φ *be a formula,* M *be a model for CMPCs,* α *be an assignment and* w *be a world in* M*. Then* $M, \alpha, w \models \varphi[y/x]$ *iff* $M, \alpha(x|\alpha(y)), w \models \varphi$.

Proof (Sketch). By induction on φ, show that for any assignment α and any world w, the following two equivalences hold: $M, \alpha, w \models \varphi[y/x]$ iff $M, \alpha(x|\alpha(y)), w \models \varphi$ and $\varphi[y/x]$ is an α_w-formula iff φ is an $\alpha(x|\alpha(y))_w$-formula. □

Here we import Seligman [14]'s propositions on validity into our semantic setting. The former explains why CMPCs must give up axiom schema K.

Proposition 3. $\Box(\varphi \supset \psi) \supset (\Box\varphi \supset \Box\psi)$ *is not valid in the class of all frames for CMPCs.*

Proof. Consider a formula $\Box(Px \supset \exists xPx) \supset (\Box Px \supset \Box\exists xPx)$. Let M be a model $(W, \{D_w\}_{w\in W}, R, V)$ where $W = \{w, v\}$, $D_w = \{a\}$, $D_v = \{b\}$, $R = \{(w, v)\}$, $V_w(Q) = V_v(Q) = \varnothing$ for all predicate symbols Q, and α be the assignment defined by $\alpha(y) = a$ for all variables y. Then $M, \alpha, w \models \Box(Px \supset \exists xPx)$ and $M, \alpha, w \models \Box Px$, but $M, \alpha, w \not\models \Box\exists xPx$. □

Proposition 4. *If* $\mathsf{FV}(\varphi) \subseteq \mathsf{FV}(\psi)$, $\Box(\varphi \supset \psi) \supset (\Box\varphi \supset \Box\psi)$ *is valid in the class of all frames for CMPCs.*

Proof. Suppose $\mathsf{FV}(\varphi) \subseteq \mathsf{FV}(\psi)$ and take any model M based on any frame, world w, assignment α such that $\Box(\varphi \supset \psi) \supset (\Box\varphi \supset \Box\psi)$ is an α_w-formula. Assume $M, \alpha, w \models \Box(\varphi \supset \psi)$ and $M, \alpha, w \models \Box\varphi$. Our goal is to show $M, \alpha, w \models \Box\psi$, so take any world v such that wRv and ψ is an α_v-formula. Then, φ is also an α_v-formula as $\mathsf{FV}(\varphi) \subseteq \mathsf{FV}(\psi)$. Hence we have $M, \alpha, v \models \varphi \supset \psi$ and $M, \alpha, v \models \varphi$ so $M, \alpha, v \models \psi$, as required. □

Proposition 5. $(\Box\varphi \wedge \Box\psi) \supset \Box(\varphi \wedge \psi)$ *is valid in the class of all frames for CMPCs, but* $\Box(\varphi \wedge \psi) \supset (\Box\varphi \wedge \Box\psi)$ *is not.*

Proof. To prove the former is easy, so we show only the latter. Consider a formula $\Box(Px \wedge \exists xPx) \supset (\Box Px \wedge \Box\exists xPx)$, and take the same model M and assignment α as those defined in the proof of Proposition 3. Then $M, \alpha, w \models \Box(Px \wedge \exists xPx)$, but $M, \alpha, w \not\models \Box\exists xPx$. □

Definition 6 (Frame properties). *Let* $F = (W, \{D_w\}_{w\in W}, R)$ *be a frame for CMPCs.*

1 F *is* serial with n objects *if for any world* $w \in W$ *and objects* $d_1, \ldots, d_n \in D_w$*, there is a world* v *such that* wRv *and* $d_1, \ldots, d_n \in D_v$.
2 F *is* reflexive *if for any world* $w \in W$, wRw.

Note that seriality with 0 *object is the ordinary seriality and that seriality with n objects implies seriality with m objects for $m \leqslant n$.*

Example 1. *Consider a world w in which agents a, b, c are drowning in a river. In this case, seriality with* 2 *objects guarantees that w has at least six worlds $v_1, \ldots, v_6$ such that up to two agents are still alive. Figure 2 presents such a scenario.*

Fig. 2. A scenario guaranteed by seriality with 2 objects

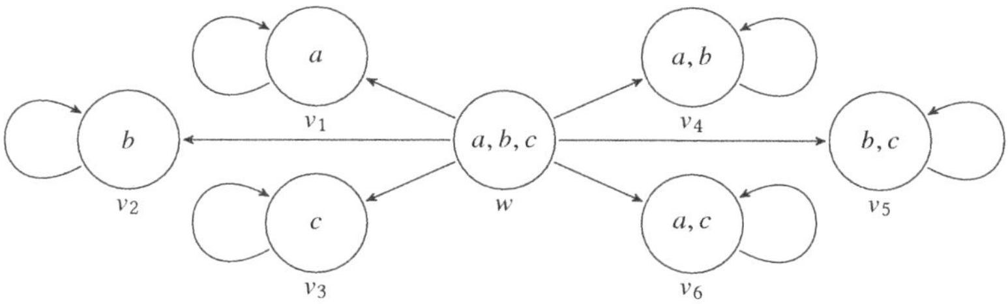

Definition 7. *A set Γ of formulas* defines *a class $\mathbb{F}$ of frames for CMPCs when the equivalence $F \models \Gamma$ iff $F \in \mathbb{F}$ holds for any frame F. If $\Gamma = \{ \varphi \}$, we say that φ defines $\mathbb{F}$.*

Proposition 8. *If Γ_i defines $\mathbb{F}_i$ for each $i \in I$, then $\bigcup_{i \in I} \Gamma_i$ defines $\bigcap_{i \in I} \mathbb{F}_i$.*

Proposition 9 (Frame definability).

1 $\mathrm{D}_n := \Box P x_1 \ldots x_n \supset \Diamond P x_1 \ldots x_n$ *defines seriality with n objects.*
2 $\mathrm{T} := \Box P \supset P$ *defines reflexivity.*

Proof. We show only that D_n defines seriality with n objects. For the right-to-left direction, take any frame F such that F is serial with n objects. Take also any valuation V, world w, assignment α such that $\Box P x_1 \ldots x_n \supset \Diamond P x_1 \ldots x_n$ is an α_w-formula. Suppose $(F, V), \alpha, w \models \Box P x_1 \ldots x_n$. By seriality with n objects of F, we have some world v such that wRv and $\alpha(x_1), \ldots, \alpha(x_n) \in D_v$. As $\alpha(x_1), \ldots, \alpha(x_n) \in D_v$ implies that $P x_1 \ldots x_n$ is an α_v-formula, we get $(F, V), \alpha, w \models \Diamond P x_1 \ldots x_n$. For the left-to-right direction, take any frame F such that $F \models \Box P x_1 \ldots x_n \supset \Diamond P x_1 \ldots x_n$, any world w, and any objects $d_1, \ldots, d_n \in D_w$. Define a valuation V and an assignment α such that $V_u(P) = D_u^n$ for any world u and $\alpha(x_i) = d_i$. Then $(F, V), \alpha, w \models \Box P x_1 \ldots x_n$. Thus we get $(F, V), \alpha, w \models \Diamond P x_1 \ldots x_n$, which implies the existence of some world v such that wRv and $d_1, \ldots, d_n \in D_v$. □

Note that we can also establish by Proposition 8 that $\mathrm{D} := \{ \mathrm{D}_n \mid n \in \mathbb{N} \}$ defines seriality with n objects for all $n \in \mathbb{N}$.

2 Hilbert-style systems for CMPCs

The Hilbert-style system $\mathrm{H}(\mathbf{K})$ for the minimal common sense modal predicate calculus $\mathbf{K}$ consists of axiom schemata and inference rules of Table 1. For the other Hilbert-style systems, we set $\Xi := \{ \mathrm{T} \} \cup \{ \mathrm{D}_n \mid n \in \mathbb{N} \}$ in what follows.

Table 1. Hilbert-style system $\mathsf{H}(\mathbf{K})$ for the minimal CMPC **K**

Axiom schemata		
A1	$\varphi \supset (\psi \supset \varphi)$	
A2	$(\varphi \supset (\psi \supset \gamma)) \supset ((\varphi \supset \psi) \supset (\varphi \supset \gamma))$	
A3	$(\neg\psi \supset \neg\varphi) \supset (\varphi \supset \psi)$	
U	$\forall x\varphi \supset \varphi[y/x]$	
K_{inv}	$\Box(\varphi \supset \psi) \supset (\Box\varphi \supset \Box\psi)$	if $\mathsf{FV}(\varphi) \subseteq \mathsf{FV}(\psi)$
Inference rules		
MP	From $\varphi \supset \psi$ and φ, we may infer ψ	
G	From $\varphi \supset \psi[y/x]$, we may infer $\varphi \supset \forall x\psi$	if y is not a free variable in $\varphi, \forall x\psi$
N	From φ, we may infer $\Box\varphi$	

Definition 10. *Given $\Sigma \subseteq \Xi$, the other Hilbert-style systems $\mathsf{H}(\mathbf{K}\Sigma)$ are obtained from $\mathsf{H}(\mathbf{K})$ by adding axiom schemata of all formulas in Σ, where axiom schema of D_n is $\Box\varphi \supset \Diamond\varphi$ such that the size of $\mathsf{FV}(\varphi)$ is at most n.*

Example 2. *If we read $\Box\varphi$ as "it is morally obligatory that φ," we could understand axiom schema of D_n as a statement that there are no moral dilemmas* for at most n agents. *Consider again the world w in which agents a, b, c are drowning in the river. In this case, D_2 guarantees that there are no moral dilemmas for at most 2 agents. However, it does not guarantee that there are no moral dilemmas for 3 agents.*

We define the notion of proof in a Hilbert-style system as usual and denote by $\vdash_{\mathsf{H}(\mathbf{K}\Sigma)} \varphi$ that a formula φ is provable in $\mathsf{H}(\mathbf{K}\Sigma)$. Some of well known theorems in the minimal normal modal predicate logic are also theorems in $\mathsf{H}(\mathbf{K})$.

Proposition 11. $\vdash_{\mathsf{H}(\mathbf{K})} (\Box\varphi_1 \wedge \cdots \wedge \Box\varphi_n) \supset \Box(\varphi_1 \wedge \cdots \wedge \varphi_n)$.

Proof. We prove only for the case $n = 2$.

1. $\vdash_{\mathsf{H}(\mathbf{K})} \varphi_1 \supset (\varphi_2 \supset (\varphi_1 \wedge \varphi_2))$ — Tautology
2. $\vdash_{\mathsf{H}(\mathbf{K})} \Box\varphi_1 \supset \Box(\varphi_2 \supset (\varphi_1 \wedge \varphi_2))$ — 1, N, K_{inv}, $\mathsf{FV}(\varphi_1) \subseteq \mathsf{FV}(\varphi_2 \supset (\varphi_1 \wedge \varphi_2))$
3. $\vdash_{\mathsf{H}(\mathbf{K})} \Box(\varphi_2 \supset (\varphi_1 \wedge \varphi_2)) \supset (\Box\varphi_2 \supset \Box(\varphi_1 \wedge \varphi_2))$ — K_{inv}, $\mathsf{FV}(\varphi_2) \subseteq \mathsf{FV}(\varphi_1 \wedge \varphi_2)$
4. $\vdash_{\mathsf{H}(\mathbf{K})} (\Box\varphi_1 \wedge \Box\varphi_2) \supset \Box(\varphi_1 \wedge \varphi_2)$ — 2, 3, Propositional Logic

□

Proposition 12. $\vdash_{\mathsf{H}(\mathbf{K})} \Diamond\neg(\varphi \supset \varphi) \supset\subset \bot$.

Proof.

1. $\vdash_{\mathsf{H}(\mathbf{K})} \neg\neg(\varphi \supset \varphi)$ — Tautology
2. $\vdash_{\mathsf{H}(\mathbf{K})} \Box\neg\neg(\varphi \supset \varphi)$ — 1, N
3. $\vdash_{\mathsf{H}(\mathbf{K})} \Diamond\neg(\varphi \supset \varphi) \supset \bot$ — 2, Propositional Logic
4. $\vdash_{\mathsf{H}(\mathbf{K})} \bot \supset \Diamond\neg(\varphi \supset \varphi)$ — Tautology
5. $\vdash_{\mathsf{H}(\mathbf{K})} \Diamond\neg(\varphi \supset \varphi) \supset\subset \bot$ — 3, 4, Propositional Logic

□

Proposition 13 (Soundness). *Let φ be a formula and $\Sigma \subseteq \Xi$. If φ is provable in $\mathsf{H}(\mathbf{K}\Sigma)$, then φ is valid in the class of all frames defined by Σ.*

Proof. By induction on height of proofs of φ in $\mathsf{H}(\mathbf{K}\Sigma)$. The validity of axiom schema K_{inv} is shown in Proposition 4 and the validities of axiom schemata T and D_n are similarly given as in Proposition 9. □

3 Strong Completeness for CMPCs

In this section, we use the canonical model construction to prove strong completeness of Hilbert-style systems for CMPCs.

Let $\Sigma \subseteq \Xi = \{\mathrm{T}\} \cup \{\mathrm{D}_n \mid n \in \mathbb{N}\}$ and $\Lambda := \mathbf{K}\Sigma$ throughout this section. We say that a formula φ is *provable from a set* Γ *of formulas in* $\mathsf{H}(\Lambda)$, denoted by $\Gamma \vdash_{\mathsf{H}(\Lambda)} \varphi$, if there exists a finite subset Δ of Γ such that $\vdash_{\mathsf{H}(\Lambda)} \bigwedge \Delta \supset \varphi$, where $\bigwedge \Delta$ denotes the conjunction of all formulas in Δ ($\bigwedge \varnothing := \top$). We also say that Γ is Λ-*inconsistent* if $\Gamma \vdash_{\mathsf{H}(\Lambda)} \bot$; Γ is Λ-*consistent* if Γ is not Λ-inconsistent; Γ is a *maximally* Λ-*consistent* set (Λ-MCS for short) if Γ is Λ-consistent and $\varphi \in \Gamma$ or $\neg\varphi \in \Gamma$ for any formula φ in $\mathsf{Form}(\Gamma) = \{\psi \in \mathsf{Form} \mid \mathsf{FV}(\psi) \subseteq \mathsf{FV}(\Gamma)\}$; Γ is *witnessed* if $\neg\forall x\varphi \in \Gamma$ implies that there is a variable $y \in \mathsf{FV}(\Gamma)$ such that $\neg\varphi[y/x] \in \Gamma$. We set $\mathcal{L}^+ := \mathcal{L}(\mathsf{Var}^+)$ and $\mathsf{Var}^+ := \mathsf{Var} \cup \mathsf{Var}'$, where Var' is a countably infinite set of variables disjoint from Var.

Proposition 14. *Let Γ be a Λ-MCS in $\mathcal{L}^+$. If $\Gamma \vdash_{\mathsf{H}(\Lambda)} \varphi$ then $\varphi \in \Gamma$ for any $\varphi \in \mathsf{Form}(\Gamma)$.*

Lemma 2 (Lindenbaum Lemma). *Let X be a countably infinite subset of Var^+ and Γ be a Λ-consistent set of formulas in $\mathcal{L}^+$ such that $X \setminus \mathsf{FV}(\Gamma)$ is infinite. There is a witnessed Λ-MCS Γ^+ in $\mathcal{L}^+$ such that $\Gamma \subseteq \Gamma^+$, $X \setminus \mathsf{FV}(\Gamma^+)$ is infinite and $\mathsf{FV}(\Gamma^+) \subseteq \mathsf{FV}(\Gamma) \cup X$.*

Definition 15 (Canonical model, Seligman [14]). *The* canonical Λ-model *is a tuple $M^\Lambda = (W^\Lambda, \{D_\Gamma\}_{\Gamma \in W^\Lambda}, R^\Lambda, V^\Lambda)$ in $\mathcal{L}^+$, where*

- $W^\Lambda = \{\Gamma \mid \Gamma$ *is a witnessed* Λ-*MCS such that* $\mathsf{Var}^+ \setminus \mathsf{FV}(\Gamma)$ *is infinite*$\}$;
- $\Gamma R^\Lambda \Delta$ *iff* $\Box\varphi \in \Gamma$ *implies* $\varphi \in \Delta$ *for any formula* φ *in* $\mathsf{Form}(\Delta)$;
- $D_\Gamma = \mathsf{FV}(\Gamma)$;
- $(x_1, \ldots, x_n) \in V^\Lambda_\Gamma(P)$ *iff* $Px_1 \ldots x_n \in \Gamma$.

The canonical assignment *is the assignment defined by* $\alpha(x) = x$.

Lemma 3 (Truth Lemma, Seligman [14]). *Let M^Λ be the canonical Λ-model and α be the canonical assignment in $\mathcal{L}^+$. For any formula φ and any $\Gamma \in W^\Lambda$ such that $\varphi \in \mathsf{Form}(\Gamma)$,*

$$\varphi \in \Gamma \quad \textit{iff} \quad M^\Lambda, \alpha, \Gamma \models \varphi.$$

Proof. Most of cases are routine and the case that φ is of the form $\forall x\psi$ is justified by Lemma 1. We prove only the right-to-left direction when φ is of the form $\Box\psi$. Suppose $\Box\psi \notin \Gamma$. We show that there is some $\Delta \in W^\Lambda$ such that $\Gamma R^\Lambda \Delta$, ψ is an α_Δ-formula and $M^\Lambda, \alpha, \Delta \not\models \psi$. Let $\Delta_0 := \{\neg\psi\} \cup \{\gamma \mid \mathsf{FV}(\gamma) \subseteq \mathsf{FV}(\psi) \text{ and } \Box\gamma \in \Gamma\}$. To apply Lindenbaum Lemma to Δ_0, we first prove that Δ_0 is Λ-consistent. Suppose for contradiction that Δ_0 is Λ-inconsistent. Thereby we have $\vdash_{\mathsf{H}(\Lambda)} (\gamma_1 \wedge \cdots \wedge \gamma_n \wedge \neg\psi) \supset \bot$ for some $\{\gamma_1, \ldots, \gamma_n\} \subseteq \Delta_0$, so we get $\Gamma \vdash_{\mathsf{H}(\Lambda)} \Box\psi$ as follows.

1	$\vdash_{\mathsf{H}(\Lambda)} (\gamma_1 \wedge \cdots \wedge \gamma_n) \supset \psi$	Assumption
2	$\vdash_{\mathsf{H}(\Lambda)} \Box(\gamma_1 \wedge \cdots \wedge \gamma_n) \supset \Box\psi$	N, K_{inv}, $\mathsf{FV}(\gamma_i) \subseteq \mathsf{FV}(\psi)$
3	$\vdash_{\mathsf{H}(\Lambda)} (\Box\gamma_1 \wedge \cdots \wedge \Box\gamma_n) \supset \Box(\gamma_1 \wedge \cdots \wedge \gamma_n)$	Proposition 11
4	$\Gamma \vdash_{\mathsf{H}(\Lambda)} \Box\psi$	2, 3, $\Box\gamma_i \in \Gamma$, Propositional Logic

However, $\Box\psi$ is in $\mathsf{Form}(\Gamma)$ so by Proposition 14 we have $\Box\psi \in \Gamma$, which contradicts our initial supposition. Hence Δ_0 is Λ-consistent. Set $X := \mathsf{Var}^+ \backslash \mathsf{FV}(\Gamma)$. Then $X \backslash \mathsf{FV}(\Delta_0)$ is infinite, so by Lindenbaum Lemma we can construct a witnessed Λ-MCS Δ such that $\Delta_0 \subseteq \Delta$, $X \backslash \mathsf{FV}(\Delta)$ is infinite, and $\mathsf{FV}(\Delta) \subseteq \mathsf{FV}(\Delta_0) \cup X$. What we need to establish is that $\Gamma R^\Lambda \Delta$, ψ is an α_Δ-formula and $M^\Lambda, \alpha, \Delta \not\models \psi$. We show only that $\Gamma R^\Lambda \Delta$ because the others are immediately established. Take any $\gamma \in \mathsf{Form}(\Delta)$ such that $\Box\gamma \in \Gamma$. To show $\gamma \in \Delta$, it suffices to show $\mathsf{FV}(\gamma) \subseteq \mathsf{FV}(\psi)$, which is justified by two claims that $\mathsf{FV}(\gamma) \subseteq \mathsf{FV}(\Delta) \cap \mathsf{FV}(\Gamma)$ and that $\mathsf{FV}(\Delta) \cap \mathsf{FV}(\Gamma) \subseteq \mathsf{FV}(\psi)$. The former is trivial and the latter holds since $\mathsf{FV}(\Delta) \subseteq \mathsf{FV}(\Delta_0) \cup X$, $X = \mathsf{Var}^+ \backslash \mathsf{FV}(\Gamma)$ and $\mathsf{FV}(\Delta_0) = \mathsf{FV}(\psi)$. □

Proposition 16. *Let M^Λ be the canonical Λ-model, where $\Lambda = \mathbf{K}\Sigma$ and $\Sigma \subseteq \Xi$. The frame F^Λ of M^Λ satisfies the frame property defined by Σ.*

Proof. By Proposition 8, it suffices to show that F^Λ satisfies the frame property defined by φ for each formula $\varphi \in \Sigma$. It is easy to show when φ is T, so we consider when φ is D_n. What we should prove is that F^Λ satisfies seriality with n objects. Take any world $\Gamma \in W^\Lambda$ and objects $x_1, \ldots, x_n \in D_\Gamma$. Let $\Delta_0 := \{\, \gamma \mid \mathsf{FV}(\gamma) \subseteq \{\, x_1, \ldots, x_n \,\} \text{ and } \Box\gamma \in \Gamma \,\}$. To apply Lindenbaum Lemma to Δ_0, similarly as in proof of Lemma 3, we first establish that Δ_0 is Λ-consistent. Suppose for contradiction that Δ_0 is Λ-inconsistent. Thereby we have $\vdash_{\mathsf{H}(\Lambda)} (\gamma_1 \wedge \cdots \wedge \gamma_m) \supset \bot$ for some $\{\, \gamma_1, \cdots, \gamma_m \,\} \subseteq \Delta_0$, so we get $\Gamma \vdash_{\mathsf{H}(\Lambda)} \Diamond\neg(Px_1 \ldots x_n \supset Px_1 \ldots x_n)$ as follows.

1	$\vdash_{\mathsf{H}(\Lambda)} (\gamma_1 \wedge \cdots \wedge \gamma_m) \supset \bot$	Assumption
2	$\vdash_{\mathsf{H}(\Lambda)} \bot \supset \neg(Px_1 \ldots x_n \supset Px_1 \ldots x_n)$	Tautology
3	$\vdash_{\mathsf{H}(\Lambda)} (\gamma_1 \wedge \cdots \wedge \gamma_m) \supset \neg(Px_1 \ldots x_n \supset Px_1 \ldots x_n)$	1, 2, Propositional Logic
4	$\vdash_{\mathsf{H}(\Lambda)} \Box(\gamma_1 \wedge \cdots \wedge \gamma_m) \supset \Box\neg(Px_1 \ldots x_n \supset Px_1 \ldots x_n)$	3, N, K_{inv}, $\mathsf{FV}(\gamma_i) \subseteq \{x_1, \ldots, x_n\}$
5	$\vdash_{\mathsf{H}(\Lambda)} \Box\neg(Px_1 \ldots x_n \supset Px_1 \ldots x_n) \supset \Diamond\neg(Px_1 \ldots x_n \supset Px_1 \ldots x_n)$	D_n
6	$\vdash_{\mathsf{H}(\Lambda)} \Box(\gamma_1 \wedge \cdots \wedge \gamma_m) \supset \Diamond\neg(Px_1 \ldots x_n \supset Px_1 \ldots x_n)$	4, 5, Propositional Logic
7	$\vdash_{\mathsf{H}(\Lambda)} (\Box\gamma_1 \wedge \cdots \wedge \Box\gamma_m) \supset \Box(\gamma_1 \wedge \cdots \wedge \gamma_m)$	Proposition 11
8	$\vdash_{\mathsf{H}(\Lambda)} (\Box\gamma_1 \wedge \cdots \wedge \Box\gamma_m) \supset \Diamond\neg(Px_1 \ldots x_n \supset Px_1 \ldots x_n)$	6, 7, Propositional Logic
9	$\Gamma \vdash_{\mathsf{H}(\Lambda)} \Diamond\neg(Px_1 \ldots x_n \supset Px_1 \ldots x_n)$	8, $\Box\gamma_i \in \Gamma$

By Proposition 12, $\Diamond\neg(Px_1 \ldots x_n \supset Px_1 \ldots x_n)$ is equivalent to $\bot$ so Γ is Λ-inconsistent, which means a contradiction. Hence Δ_0 is Λ-consistent. Similarly again as in proof of Lemma 3, let $X := \mathsf{Var}^+ \backslash \mathsf{FV}(\Gamma)$. Then $X \backslash \mathsf{FV}(\Delta_0)$ is infinite, so by Lindenbaum Lemma we can construct a witnessed Λ-MCS Δ such that $\Delta_0 \subseteq \Delta$, $X \backslash \mathsf{FV}(\Delta)$ is infinite, and $\mathsf{FV}(\Delta) \subseteq \mathsf{FV}(\Delta_0) \cup X$. By the definition of Δ_0, this Δ satisfies that $\Gamma R^\Lambda \Delta$ and $x_1, \ldots, x_n \in D_\Delta$. □

Theorem 1 (Strong completeness). *Let $\Gamma \cup \{\, \varphi \,\}$ be a set of formulas and $\Sigma \subseteq \Xi$. If φ is a consequence from Γ in the class of all frames defined by Σ, then φ is provable from Γ in $\mathsf{H}(\mathbf{K}\Sigma)$.*

4 Sequent Calculi for CMPCs

Sequent calculi for normal modal logics are well known in the literature. For example, as H. Wansing surveyed in [17], sequent calculi for the minimal normal modal logic and its extensions with axiom schema D, T are provided in [9,13,10], [5] and [11], respectively. What makes our sequent calculi for CMPCs fundamentally different from those calculi is the restriction of modal rules on variables except for T-rule.

Given finite multisets Γ, Δ of formulas, we call an expression $\Gamma \rightarrow \Delta$ a *sequent* which is intuitively read as "if all formulas in Γ are true then some formulas in Δ are true." We also denote $\{\Box\varphi \mid \varphi \in \Gamma\}$ by $\Box\Gamma$. A sequent calculus G for first-order logic consists of initial sequents, structural rules and logical rules displayed in Table 2. The sequent calculus $\mathrm{G}(\mathbf{K}\Sigma)$ is then defined as follows. Given $\Sigma \subseteq \Xi$, $\mathrm{G}(\mathbf{K}\Sigma)$ is obtained from G by adding additional logical rules $(\Box_{\mathrm{K}_{inv}})$ and $(\Box_X)$ for each X in Σ which are displayed in Table 2. Also $\mathrm{G}^-(\mathbf{K}\Sigma)$ is the calculus obtained by removing structural rule (Cut) from $\mathrm{G}(\mathbf{K}\Sigma)$ and $\mathrm{G}^*(\mathbf{K}\Sigma)$ is the calculus obtained by replacing (Cut) of $\mathrm{G}(\mathbf{K}\Sigma)$ with the extended rule (Cut^*) which is introduced in [12,8]:

$$\frac{\Gamma \rightarrow \Delta, \varphi^m \qquad \varphi^n, \Theta \rightarrow \Sigma}{\Gamma, \Theta \rightarrow \Delta, \Sigma}\ (Cut^*),$$

where m, n can be zero and each φ is called *cut-formula*. We define the notion of derivation in a sequent calculus as usual and denote by $\vdash_{\mathrm{G}(\mathbf{K}\Sigma)} \Gamma \rightarrow \Delta$ that a sequent $\Gamma \rightarrow \Delta$ is derivable in $\mathrm{G}(\mathbf{K}\Sigma)$.

Example 3. *The following is a derivation of* $\rightarrow \Box(Px \wedge Py) \supset \Diamond(Px \wedge Py)$ *in* $\mathrm{G}(\mathbf{KD}_2)$.

$$\dfrac{\dfrac{\dfrac{\dfrac{Px \wedge Py \rightarrow Px \wedge Py}{\neg(Px \wedge Py), Px \wedge Py \rightarrow}\ \neg\rightarrow}{\Box\neg(Px \wedge Py), \Box(Px \wedge Py) \rightarrow}\ \Box_{\mathrm{D}_2}}{\Box(Px \wedge Py) \rightarrow \Diamond(Px \wedge Py)}\ \rightarrow\neg}{\rightarrow \Box(Px \wedge Py) \supset \Diamond(Px \wedge Py)}\ \rightarrow\supset$$

Proposition 17 (Equipollence). *Let* $\Sigma \subseteq \Xi$. *A formula* φ *is provable in* $\mathrm{H}(\mathbf{K}\Sigma)$ *iff the sequent* $\rightarrow \varphi$ *is derivable in* $\mathrm{G}(\mathbf{K}\Sigma)$.

Proof (Sketch). The left-to-right direction is proved by induction on height of proofs of φ in $\mathrm{H}(\mathbf{K}\Sigma)$. The right-to-left direction is proved by establishing two claims: that if $\vdash_{\mathrm{G}(\mathbf{K}\Sigma)} \Gamma \rightarrow \Delta$ then $\vdash_{\mathrm{H}(\mathbf{K}\Sigma)} \bigwedge\Gamma \supset \bigvee\Delta$ and that $\vdash_{\mathrm{H}(\mathbf{K}\Sigma)} (\bigwedge\varnothing \supset \bigvee\{\varphi\}) \supset \varphi$, where $\bigwedge\Gamma$ is the conjunction of all formulas in Γ ($\bigwedge\varnothing := \top$) and $\bigvee\Delta$ is the disjunction of all formulas in Δ ($\bigvee\varnothing := \bot$). □

Our main result on sequent calculi for CMPCs is cut elimination theorem. In what follows we assume that free variables and bound variables in derivations are thoroughly separated.

Theorem 2 (Cut elimination). *Let* $\Sigma \subseteq \Xi$. *If a sequent* $\Gamma \rightarrow \Delta$ *is derivable in* $\mathrm{G}(\mathbf{K}\Sigma)$, *then* $\Gamma \rightarrow \Delta$ *is also derivable in* $\mathrm{G}^-(\mathbf{K}\Sigma)$.

Table 2. Sequent calculus G and additional logical rules

Initial sequents of G	
$\varphi \rightarrow \varphi$	
Structural rules of G	
$\frac{\Gamma \rightarrow \Delta}{\Gamma \rightarrow \Delta, \varphi}$ $(\rightarrow w)$	$\frac{\Gamma \rightarrow \Delta}{\varphi, \Gamma \rightarrow \Delta}$ $(w \rightarrow)$
$\frac{\Gamma \rightarrow \Delta, \varphi, \varphi}{\Gamma \rightarrow \Delta, \varphi}$ $(\rightarrow c)$	$\frac{\varphi, \varphi, \Gamma \rightarrow \Delta}{\varphi, \Gamma \rightarrow \Delta}$ $(c \rightarrow)$
$\frac{\Gamma \rightarrow \Delta, \varphi \qquad \varphi, \Theta \rightarrow \Sigma}{\Gamma, \Theta \rightarrow \Delta, \Sigma}$ (Cut)	
Logical rules of G	
$\frac{\varphi, \Gamma \rightarrow \Delta}{\Gamma \rightarrow \Delta, \neg\varphi}$ $(\rightarrow \neg)$	$\frac{\Gamma \rightarrow \Delta, \varphi}{\neg\varphi, \Gamma \rightarrow \Delta}$ $(\neg \rightarrow)$
$\frac{\varphi, \Gamma \rightarrow \Delta, \psi}{\Gamma \rightarrow \Delta, \varphi \supset \psi}$ $(\rightarrow \supset)$	$\frac{\Gamma \rightarrow \Delta, \varphi \qquad \psi, \Theta \rightarrow \Sigma}{\varphi \supset \psi, \Gamma, \Theta \rightarrow \Delta, \Sigma}$ $(\supset \rightarrow)$
$\frac{\Gamma \rightarrow \Delta, \varphi[y/x]}{\Gamma \rightarrow \Delta, \forall x\varphi}$ $(\rightarrow \forall)^{\star}$	$\frac{\varphi[y/x], \Gamma \rightarrow \Delta}{\forall x\varphi, \Gamma \rightarrow \Delta}$ $(\forall \rightarrow)$

$\star$: y is not a free variable in $\Gamma, \Delta, \forall x\varphi$.

Additional logical rules		
$\frac{\Gamma \rightarrow \varphi}{\Box\Gamma \rightarrow \Box\varphi}$ $(\Box\mathrm{K}_{inv})^{\dagger}$	$\frac{\Gamma \rightarrow}{\Box\Gamma \rightarrow}$ $(\Box\mathrm{D}_{n})^{\ddagger}$	$\frac{\varphi, \Gamma \rightarrow \Delta}{\Box\varphi, \Gamma \rightarrow \Delta}$ $(\Box\mathrm{T})$

$\dagger$: $\mathsf{FV}(\Gamma) \subseteq \mathsf{FV}(\varphi)$.

$\ddagger$: The size of $\mathsf{FV}(\Gamma)$ is at most n.

Proof (Sketch). Since (Cut) is an instance of (Cut^*), it is sufficient to show that $\vdash_{G^*(\mathbf{K}\Sigma)} \Gamma \rightarrow \Delta$ implies $\vdash_{G^-(\mathbf{K}\Sigma)} \Gamma \rightarrow \Delta$. To show this, we say that a derivation $\mathfrak{D}$ in $G^*(\mathbf{K}\Sigma)$ is of *the* (Cut^*)*-bottom form* if the last applied rule in $\mathfrak{D}$ is (Cut^*) and there are no other applications of (Cut^*) in $\mathfrak{D}$. We also let the *weight* of a derivation of (Cut^*)-bottom form be the number of sequents occurring in $\mathfrak{D}$ except for its root. Then, given a derivation $\mathfrak{D}$ of (Cut^*)-bottom form of a sequent $\Gamma' \rightarrow \Delta'$ in $G^*(\mathbf{K}\Sigma)$, by double induction on complexity of cut-formulas and weight of $\mathfrak{D}$, we can construct a derivation of $\Gamma' \rightarrow \Delta'$ in $G^-(\mathbf{K}\Sigma)$. This procedure tells us that $\vdash_{G^*(\mathbf{K}\Sigma)} \Gamma \rightarrow \Delta$ implies $\vdash_{G^-(\mathbf{K}\Sigma)} \Gamma \rightarrow \Delta$. For example, consider the case that $\mathfrak{D}$ is of the form

$$\dfrac{\dfrac{\dfrac{\mathfrak{D}_1}{\Gamma_1' \rightarrow \varphi}}{\Box\Gamma_1' \rightarrow \Box\varphi}\,\Box K_{inv} \qquad \dfrac{\dfrac{\mathfrak{D}_2}{(\varphi)^n, \Gamma_2' \rightarrow \psi}}{(\Box\varphi)^n, \Box\Gamma_2' \rightarrow \Box\psi}\,\Box K_{inv}}{\Box\Gamma_1', \Box\Gamma_2' \rightarrow \Box\psi}\,(Cut^*),$$

where $\Gamma' = \Box\Gamma_1' \cup \Box\Gamma_2'$ and $\Delta' = \Box\psi$; n is the number of occurrences of φ and $\Box\varphi$; $\mathsf{FV}(\Gamma_1') \subseteq \mathsf{FV}(\varphi)$ and $\mathsf{FV}(\{\varphi\} \cup \Gamma_2') \subseteq \mathsf{FV}(\psi)$. Then, $\mathsf{FV}(\Gamma_1' \cup \Gamma_2') \subseteq \mathsf{FV}(\psi)$ so we can construct a derivation $\mathfrak{D}'$ of $\Gamma' \rightarrow \Delta'$ in $G^-(\mathbf{K}\Sigma)$ as follows.

$$\dfrac{\dfrac{\dfrac{\mathfrak{D}_1}{\Gamma_1' \rightarrow \varphi} \qquad \dfrac{\mathfrak{D}_2}{(\varphi)^n, \Gamma_2' \rightarrow \psi}}{\Gamma_1', \Gamma_2' \rightarrow \psi}\,(Cut^*)}{\Box\Gamma_1', \Box\Gamma_2' \rightarrow \Box\psi}\,\Box K_{inv}$$

The application of (Cut^*) in $\mathfrak{D}'$ is then innocuous because the complexity of cut-formulas is reduced. □

Conclusion

In this paper, we have provided Hilbert-style systems and cut-free sequent calculi for CMPC **K** and its extensions with axiom schemata T, D_n, and D. We have also presented definability results in CMPC.

We shall close this paper by outlining some directions for further studies. One direction would be to add constants, function symbols and equality to CMPCs. Adding function symbols to CMPCs with axiom schema D_n would require some trick to show Proposition 16, because a natural construction of the canonical model would have one whose domains D_Γ consist of terms like $f(x, y)$ in worlds Γ, but this construction does not work due to the restriction of D_n on variables.

Another direction to be pursued would be to extend CMPC **K** with axiom schemata B, 4 and 5. To begin, we should examine the corresponding frame properties to B, 4 and 5. Probably this task will be interesting in its own right. Then, we have to provide a proof of strong completeness for each logic. The second author of the paper has found that the frame of the canonical model for CMPC **K** with 4 and T does not satisfy transitivity. To extend CMPC **K** with 4 and T, thus, we need either to revise the canonical model construction so as to make it work, or take a different strategy to prove strong completeness. The step-by-step method introduced in [2, p. 223] might be applicable for CMPC **K** with 4 and T since the original CMPC is proved by this method in Seligman's second draft [16].

Acknowledgment We are extremely grateful to Jeremy Seligman for sharing his draft on CMPC and discussing it with us. We also thank Tomoyuki Yamada for helpful comments and discussions. The work of the second author was partially supported by JSPS KAKENHI Grant-in-Aid for Scientific Research (C) Grant Number 19K12113, JSPS KAKENHI Grant-in-Aid for Scientific Research (B) Grant Number 17H02258, and JSPS Core-to-Core Program (A. Advanced Research Networks).

References

1. van Benthem, J.: Modal Logic for Open Minds. CSLI Publications (2010)
2. Blackburn, P., de Rijke, M., Venema, Y.: Modal Logic. Cambridge University Press (2002), fourth printing with corrections 2010
3. Fitting, M., Mendelsohn, R.L.: First-Order Modal Logic. Kluwer Academic Publishers (1998)
4. Garson, J.: Quantification in Modal Logic. In: Gabbay, D.M., Guenthner, F. (eds.) Handbook of Philosophical Logic, vol. 3, chap. 4, pp. 267–323. Springer, second edn. (2010)
5. Goble, L.F.: Gentzen Systems for Modal Logic. Notre Dame Journal of Formal Logic **15**(3), 455–461 (Jul 1974)
6. Hazen, A.: Counterpart-Theoretic Semantics for Modal Logic. The Journal of Philosophy **76**(6), 319–338 (Jun 1979)
7. Hughes, G.E., Cresswell, M.J.: A New Introduction to Modal Logic. Routledge (1996)
8. Kashima, R.: Mathematical Logic. Asakura Publishing Co. Ltd (2009), in Japanese
9. Leivant, D.: On the Proof Theory of the Modal Logic for Arithmetic Provability. Journal of Symbolic Logic **46**(3), 531–538 (Sep 1981)
10. Mints, G.: Gentzen-type Systems and Resolution Rules Part I Propositional Logic. In: Martin-Löf, P., Mints, G. (eds.) COLOG-88, Lecture Notes in Computer Science, vol. 417, pp. 198–231. Springer Berlin Heidelberg (1990)
11. Ohnishi, M., Matsumoto, K.: Gentzen Method in Modal Calculi. Osaka Mathematical Journal **9**(2), 113–130 (1957)
12. Ono, H., Komori, Y.: Logics without the Contraction Rule. The Journal of Symbolic Logic **50**(1), 169–201 (1985)
13. Sambin, G., Valentini, S.: The Modal Logic of Provability. The Sequential Approach. Journal of Philosophical Logic **11**(3), 311–342 (Aug 1982)
14. Seligman, J.: Common Sense Modal Predicate Logic: 1st draft (Oct 2016)
15. Seligman, J.: Common Sense Modal Predicate Logic. Presentation (2017), Non-classical Modal and Predicate Logics: The 9th International Workshop on Logic and Cognition, Guangzhou, China, 4 December 2017
16. Seligman, J.: Common Sense Modal Predicate Logic: 2nd draft (Nov 2017)
17. Wansing, H.: Sequent Systems for Modal Logics. In: Gabbay, D.M., Guenthner, F. (eds.) Handbook of Philosophical Logic, vol. 8, chap. 2, pp. 61–145. Springer Netherlands (2002)

Strengthened Conditionals*

Eric Raidl[1] [0000-0001-6153-4979]

University of Tuebingen, Germany eric.raidl@uni-tuebingen.de

Abstract. This article develops a general method to transfer completeness results from a basic conditional to a defined conditional. As an example, the method is implemented for the so-called neutral conditional.

Keywords: Definable Conditionals · Neutral Conditional · Sufficient Reason · Conditional Logic · Completeness Results · Correspondence Theory.

1 Introduction

Conditionals are natural language sentences of the form 'if A, [then] C', where A is the antecedent and C the consequent.[1] Conditionals are difficult to analyse. A standard account has however emerged, the so-called possible worlds account (Lewis, 1971; Stalnaker, 1968). According to this account, roughly, a conditional $A > C$ is true in the actual world w if and only if the closest A-worlds to w are C-worlds.[2] However, recent reflections suggest that we may want to strengthen the defining clause by additional conditions. Different approaches argue for different conditions (Crupi & Iacona, 2019; Lewis, 1973a; Raidl, 2019a; Rott, 1986, 2019; Spohn, 2015). This article develops a technique to generate logics for such strengthened conditionals.

The *general problem* is this: Consider a strengthened conditional of the form

– $\varphi \rhd \psi$ in world w iff closest φ-worlds are ψ-worlds *and* X.

Suppose that X is also formulated in terms of closeness. One can then rephrase $\varphi \rhd \psi$ in the language for $>$ as $(\varphi > \psi) \wedge \chi$, where χ expresses the semantic condition X. The *main question* is this: Can we use known completeness results for $>$ to obtain completeness results for $\rhd$? The answer is yes and the paper provides a general method: Redefine $>$ in terms of $\rhd$. This *backtranslation* of $\varphi > \psi$ yields a sentence α in the language for $\rhd$. One can then (roughly) use

* This chapter is in its final form and it is not submitted to publication anywhere else. This research was funded by the *Deutsche Forschungsgemeinschaft* (DFG, German Research Foundation) under Research Unit FOR 1614, and under Germany's Excellence Strategy by the DFG Cluster of Excellence "Machine Learning: New Perspectives for Science", EXC-Number 2064/1, Project number 390727645.

[1] Sentences of the form 'A, hence/therefore C' or 'C, because/since A' are also analysed as akin to conditionals.

[2] A more general account is used throughout the article (see §2).

B. Liao and Y. N. Wáng (eds.), *Context, Conflict and Reasoning*,
Logic in Asia: Studia Logica Library,
https://doi.org/10.1007/978-981-15-7134-3_11

this backtranslation to translate axioms and rules for $>$ into axioms and rules for $\rhd$. These backtranslates provide a logic for $\rhd$.

The plan of the paper is as follows. §2 recalls basic conditional logic. §3 introduces the translation and backtranslation between conditional languages. §4 uses these to transfer soundness, completeness and correspondence results. The method developed in §3 and §4 provides essential simplifications of the more general approach taken in Raidl (2020) and additional facts are proven. §5 implements the method for the neutral conditional. The Conclusion (§6) lists other conditionals which can be analyzed similarly.

2 Basic Conditional Logics

Let me introduce the background for the following sections. Since the results here are known or analogues to famous results in modal logic, the proofs are omitted.

The alphabet of our basic conditional language is given by a fixed set of propositional variables Var, classical connectives, $\neg, \wedge, \vee, \rightarrow$, the basic conditional $>$ and the parenthesis) and (. The set of formulas is defined inductively and is denoted $\mathcal{L}_>$. $\top$ denotes any classical tautology and $\bot = \neg\top$. $\leftrightarrow$ abbreviates the biconditional for $\rightarrow$. The following is a non-exhaustive list of possible rules and axioms of conditional logic:

$$\frac{\vdash \varphi \leftrightarrow \varphi'}{\vdash (\varphi > \chi) \rightarrow (\varphi' > \chi)}\ \textsf{RCEA} \qquad \frac{\vdash \varphi \leftrightarrow \varphi'}{\vdash (\chi > \varphi) \rightarrow (\chi > \varphi')}\ \textsf{RCEC}$$

$$\frac{\vdash \varphi \rightarrow \varphi'}{\vdash (\chi > \varphi) \rightarrow (\chi > \varphi')}\ \textsf{RCM}$$

$\varphi > \top$	CN
$(\varphi > \psi) \rightarrow (\varphi > (\psi \vee \chi))$	Cm
$\varphi > \varphi$	ID
$((\varphi > \psi) \wedge (\varphi > \chi)) \rightarrow (\varphi > (\psi \wedge \chi))$	CC
$((\varphi > \chi) \wedge (\psi > \chi)) \rightarrow ((\varphi \vee \psi) > \chi)$	CA
$((\varphi > \chi) \wedge (\varphi > \psi)) \rightarrow ((\varphi \wedge \psi) > \chi)$	CMon
$((\varphi > \chi) \wedge \neg(\varphi > \neg\psi)) \rightarrow ((\varphi \wedge \psi) > \chi)$	CV
$((\varphi > \psi) \wedge ((\varphi \wedge \psi) > \chi)) \rightarrow (\varphi > \chi)$	Cut
$((\varphi > \psi) \wedge \neg(\varphi > \bot)) \rightarrow \neg((\varphi \wedge \psi) > \bot)$	WCut
$(\varphi > \bot) \rightarrow ((\varphi \wedge \psi) > \bot)$	M
$((\varphi > \bot) \wedge (\psi > \bot)) \rightarrow ((\varphi \vee \psi) > \bot)$	C
$\neg(\varphi > \bot)$	Con

Rules and axioms are in the form presented by Chellas (1975). If available, I use his abbreviations. Cm, M, C and WCut are my abbreviations and Cut is an implicational reformulation of the KLM (Kraus, Lehmann, & Magidor, 1990) rule Cut. The above rules and axioms have analogues in the KLM framework

(Arló-Costa & Shapiro, 1992).[3] CN says that all conditionals with a classical tautology in the consequent hold. The following are known or clear: RCM is equivalent to the combination of RCEC and Cm. CN follows from RCEC, Cm and ID. M follows from RCM and CMon. C follows from CA. WCut follows from Cut. The axiom Con interdicts impossible consequent conditionals and is the sign of non-vacuist conditional logics, as we will see.

The *outer necessity* is $\Box\varphi \equiv (\neg\varphi > \bot)$. The outer possibility is defined as dual to $\Box$. M and C are then equivalent to the following outer-modality reformulations:

$$\Box\varphi \rightarrow \Box(\varphi \vee \psi) \qquad \mathsf{M}'$$
$$(\Box\varphi \wedge \Box\psi) \rightarrow \Box(\varphi \wedge \psi) \qquad \mathsf{C}'$$

In other words, M expresses that the outer necessity is monotone and C expresses that the outer necessity is closed under conjunction.

$\mathsf{L} \subseteq \mathcal{L}_>$ is a *conditional logic* iff it contains all substitution instances of propositional tautologies (PT) and is closed under Modus Ponens for $\rightarrow$ (MoPo). A conditional logic L is *classical* iff it is also closed under RCEA and RCEC, *consequent-monotone*[4] iff it is closed under RCEA and RCM and *normal* iff it is consequent-monotone and contains the axioms CN and CC. The smallest classical, consequent-monotone and normal conditional logics are respectively denoted CE, CM and CK. For KLM-scholars, note that throughout the article CM stands for consequent-monotone conditional logic (defined above) and *not* for cautious monotonicity, here denoted CMon. In a classical conditional logic, substitution of provable equivalents is derivable. We denote $\mathsf{L} + \mathsf{X}_1 + \ldots + \mathsf{X}_n$ the smallest conditional logic closed under the rules of L, containing the axioms of L, as well as the axioms $\mathsf{X}_1, \ldots, \mathsf{X}_n$. As an example, CM and CK can alternatively be characterised by $\mathsf{CM} = \mathsf{CE} + \mathsf{Cm}$, $\mathsf{CK} = \mathsf{CM} + \mathsf{CN} + \mathsf{CC}$. Furthermore, we denote $\mathsf{CU} = \mathsf{CM} + \mathsf{ID} + \mathsf{Cut} + \mathsf{CMon}$ the (full) cumulative logic,[5] $\mathsf{P} = \mathsf{CU} + \mathsf{CA}$ preferential logic, and Lewis's (1971) weakest conditional logic is $\mathsf{V} = \mathsf{CK} + \mathsf{ID} + \mathsf{CA} + \mathsf{CMon} + \mathsf{CV}$. The non-nested fragment of P and V correspond respectively to the system **P** and **R** of Kraus et al. (1990).[6]

In what follows, $f\colon X \longrightarrow Y$ indicates that f is a total function from X to Y. To model the logic CE and extensions, I adopt Chellas' (1975) flexible semantics:

Definition 1. Let W be a non-empty set. $\mathfrak{F} = \langle W, F \rangle$ is a *minimal frame* iff $F\colon (W \times \wp(W)) \longrightarrow \wp(\wp(W))$. $\mathfrak{M} = \langle W, F, V \rangle$ is a *minimal model* for $\mathcal{L}_>$ iff $\langle W, F \rangle$ is a minimal frame and $V\colon \mathrm{Var} \longrightarrow \wp(W)$.

---

[3] Restrict φ, ψ, χ to conditional free formulas, replace the connective $>$ by the inference relation $\mid\!\sim$ and restate everything in a rule form, where $\wedge$, $\rightarrow$ and $\neg$ having conditionals in their scope are reformulated in the meta-language. As an example, CV is: If $\varphi \mid\!\sim \chi$ and $\varphi \not\mid\!\sim \neg\psi$ then $\varphi \wedge \psi \mid\!\sim \chi$. Here are the most well known KLM abbreviations: LLE for RCEA, RLE for RCEC, RW for RCM, Refl for ID, AND for CC, OR for CA, CM for CMon, RM for CV, Cut for Cut.

[4] Chellas calls this "monotone conditional logic".

[5] CC is derivable here.

[6] Arló-Costa and Shapiro (1992) show that the thesis of **R** can be mapped to the generalised Horn fragment of V.

Points in W are *worlds*, subsets of W are *propositions*. The *neighbourhood selection function* F associates to every world and every proposition A a set of propositions $F(w, A)$ – the A-*neighbourhood* to w. One may think of it as the set of options triggered by supposing A, or as the set of believed propositions after revising by A. The above semantics is to Lewis-Stalnaker semantics of set-selection models (coined in the introduction) what neighbourhood semantics is to Kripke semantics.

Definition 2. *Truth* in a minimal model $\mathfrak{M}$ for the language $\mathcal{L}_>$ is denoted $\models_>^{\mathfrak{M}}$. Truth for propositional variables and classical connectives is defined as usual (in modal logic), and

(>) $w \models_>^{\mathfrak{M}} \varphi > \psi$ iff $[\psi]_>^{\mathfrak{M}} \in F(w, [\varphi]_>^{\mathfrak{M}})$.

Here $[\varphi]_>^{\mathfrak{M}} = \{w \in W : w \models_>^{\mathfrak{M}} \varphi\}$ are co-inductively defined.

We write $\mathfrak{M} \models_> \varphi$ iff for all $w \in W, w \models_>^{\mathfrak{M}} \varphi$. $\mathfrak{F} \models_> \varphi$ iff for all models $\mathfrak{M}$ over $\mathfrak{F}$, $\mathfrak{M} \models_> \varphi$. $C \models_> \varphi$ (for C a class of models or frames) iff for all $X \in C$, $X \models_> \varphi$. When the model is clear from the context, we drop upper indices, writing $\models_>$ instead of $\models_>^{\mathfrak{M}}$ and $[\varphi]_>$ instead of $[\varphi]_>^{\mathfrak{M}}$.

By a canonical model construction, Chellas (1975) proved:

Theorem 1. CE *is sound and complete for minimal frames.*

We can *semanticise* axioms X into corresponding properties of F: for φ becoming $A = [\varphi]$, ψ becoming B, and χ becoming C, $\varphi > \psi$ becomes $B \in F(w, A)$. Inner operators (in the scope of >) are translated algebraically: $\top$ becomes W, $\bot$ becomes $\emptyset$, $\neg\varphi$ becomes $\overline{A} = W \setminus A$, $\wedge$ becomes $\cap$, $\vee$ becomes $\cup$ and $\varphi \to \psi$ becomes $\overline{A} \cup B$. Outer operators are translated in the natural language. For X a conditional axiom, I denote X^F, or just (x), the corresponding property for minimal frames obtained by this semantisation. For example, the frame property corresponding to CV is:

$$C \in F(w, A) \text{ and } \overline{B} \notin F(w, A) \text{ implies } C \in F(w, A \cap B) \qquad \mathsf{CV}^F, \text{cv}$$

By these remarks, the correspondence theory is straightforward (proof omitted):

Theorem 2. *For $\mathfrak{F}$ a minimal frame, $\mathfrak{F} \models \mathsf{X}$ iff $\mathfrak{F}$ satisfies X^F.*

Using Chellas' canonical model construction for Theorem 1, completeness results for extensions of CE easily follow (proof omitted):

Theorem 3. *Let $\mathsf{X}_1, \ldots, \mathsf{X}_n$ axioms from the above list. The logic $\mathsf{CE} + \mathsf{X}_1 + \ldots + \mathsf{X}_n$ is sound and complete for the class M of minimal frames (or models) with the properties $\mathsf{X}_1^F, \ldots, \mathsf{X}_n^F$.*

3 Translating

This section introduces the idea of translating between two conditional languages. We consider another conditional language $\mathcal{L}_\rhd$ which is like $\mathcal{L}_>$, except that $\rhd$ is a notational variant of $>$ which will ultimately be interpreted differently. The shared fragment of the two languages is the classical propositional language, denoted $\mathcal{L}$.

As example, we take the language $\mathcal{L}_{\gtrdot}$, where $\gtrdot$ is to be interpreted as the neutral conditional (Raidl, 2019a). The frames and models are the same. Truth clauses are not. Truth for $\mathcal{L}_{\gtrdot}$ in a minimal model $\mathfrak{M}$ is denoted $\vDash_{\gtrdot}^{\mathfrak{M}}$. The truth clauses for the propositional variables and the classical connectives are the same as for $\mathcal{L}_{>}$ and

$$(\gtrdot)\ \ w \vDash_{\gtrdot}^{\mathfrak{M}} \varphi \gtrdot \psi \text{ iff } [\psi]_{\gtrdot}^{\mathfrak{M}} \in F(w, [\varphi]_{\gtrdot}^{\mathfrak{M}}) \text{ and } [\bot]_{\gtrdot}^{\mathfrak{M}} \notin F(w, [\varphi]_{\gtrdot}^{\mathfrak{M}}). \qquad \text{(neutral)}$$

If $>$ were in the language $\mathcal{L}_{\gtrdot}$, we could define $\varphi \gtrdot \psi$ as $(\varphi > \psi) \wedge \neg(\varphi > \bot)$. However, $>$ is not in $\mathcal{L}_{\gtrdot}$. To state a relation between $\mathcal{L}_{\gtrdot}$ and $\mathcal{L}_{>}$, we thus need other resources.

Let α be a formula of $\mathcal{L}_{>}$ and p, q propositional variables. We write $\alpha = \alpha[p, q]$ iff α has its propositional variables among $\{p, q\}$. We write $\alpha[p, q] \in \mathcal{L}_{>}$ as an abbreviation for $\alpha \in \mathcal{L}_{>}, p, q \in \text{Var}$ and $\alpha = \alpha[p, q]$. Let φ, ψ formulas of $\mathcal{L}_{>}$ and $\alpha[p, q] \in \mathcal{L}_{>}$, then we write $\alpha[\varphi/p, \psi/q]$ for substituting simultaneously φ for p and ψ for q in α.

Definition 3. $\circ\colon \mathcal{L}_{\rhd} \longrightarrow \mathcal{L}_{>}$ is a *translation* iff $p^\circ = p$, for $p \in \text{Var}$, $(\neg\varphi)^\circ = \neg\varphi^\circ$, $(\varphi * \psi)^\circ = (\varphi^\circ * \psi^\circ)$ for $* \in \{\wedge, \vee, \rightarrow\}$, and there is a formula $\alpha[p, q] \in \mathcal{L}_{>}$ such that for every φ, ψ in $\mathcal{L}_{\rhd}$, $(\varphi \rhd \psi)^\circ = \alpha[\varphi^\circ/p, \psi^\circ/q]$.

Intuitively α provides the *form* for the translate of $\varphi \rhd \psi$. For our example, we can read off the form α for $(\varphi \rhd \psi)^\circ$ from the semantic definition of the defined conditional:

$$(\circ_{\gtrdot})\ \ (\varphi \gtrdot \psi)^\circ := (\varphi^\circ > \psi^\circ) \wedge \neg(\varphi^\circ > \bot)$$

We can also show that:

Lemma 1. *Let* $\circ\colon \mathcal{L}_{\rhd} \longrightarrow \mathcal{L}_{>}$ *be a translation,* $\theta = \theta[p_1, \ldots, p_n] \in \mathcal{L}$ *and* $\varphi_1, \ldots, \varphi_n \in \mathcal{L}_{\rhd}$. *Then* $(\theta[\varphi_1/p_1, \ldots, \varphi_n/p_n])^\circ = \theta[\varphi_1^\circ/p_1, \ldots, \varphi_n^\circ/p_n])$.

Proof. By induction on the complexity of the formula. Since $\theta \in \mathcal{L}$, we need only consider the classical connectives.

$\theta = p_i$ for $i \in \{1, \ldots, n\}$. Then $\theta[\varphi_1/p_1, \ldots, \varphi_n/p_n] = \varphi_i$ and $\theta[\varphi_1^\circ/p_1, \ldots, \varphi_n^\circ/p_n] = \varphi_i^\circ$. Thus $(\theta[\varphi_1/p_1, \ldots, \varphi_n/p_n])^\circ = \varphi_i^\circ = \theta[\varphi_1^\circ/p_1, \ldots, \varphi_n^\circ/p_n]$.

$\theta = \neg\gamma$. That is $\gamma = \gamma[p_1, \ldots, p_n] \in \mathcal{L}$. Assume the induction hypothesis (IH) for γ. That is $(\gamma[\varphi_1/p_1, \ldots, \varphi_n/p_n])^\circ = \gamma[\varphi_1^\circ/p_1, \ldots, \varphi_n^\circ/p_n]$. Thus

$$\begin{aligned}
(\theta[\varphi_1/p_1, \ldots, \varphi_n/p_n])^\circ &= (\neg\gamma[\varphi_1/p_1, \ldots, \varphi_n/p_n])^\circ && (\theta = \neg\gamma)\\
&= \neg(\gamma[\varphi_1/p_1, \ldots, \varphi_n/p_n])^\circ && (\circ)\\
&= \neg\gamma[\varphi_1^\circ/p_1, \ldots, \varphi_n^\circ/p_n] && (\text{IH})\\
&= \theta[\varphi_1^\circ/p_1, \ldots, \varphi_n^\circ/p_n] && (\theta = \neg\gamma)
\end{aligned}$$

$\theta = \gamma_1 \wedge \gamma_2$, $\theta = \gamma_1 \vee \gamma_2$ and $\theta = \gamma_1 \rightarrow \gamma_2$ can be treated in a similarly. $\square$

It follows that if $\varphi \in \mathcal{L}$, then $\varphi^\circ = \varphi$. Thus, since $\top \in \mathcal{L}$ and $\bot = \neg\top$, we obtain $\top^\circ = \top$ and $\bot^\circ = \bot$.

A translation maps formulas of $\mathcal{L}_\rhd$ to formulas of $\mathcal{L}_>$. It should also conserve meaning. To ensure this, we need a semantic relation between the interpreted models for $>$ and $\rhd$.

Definition 4. Let $\circ\colon \mathcal{L}_\rhd \longrightarrow \mathcal{L}_>$ be a translation and M a model class in $\mathcal{L}_>$. M *is* $\circ$*-automorphic to itself (on the right)*, denoted $M \overset{\circ}{\approx} M$, iff for all $\varphi \in \mathcal{L}_\rhd$, all $\mathfrak{M} \in M$ and all $w \in W$, we have: $w \vDash_\rhd^{\mathfrak{M}} \varphi$ iff $w \vDash_>^{\mathfrak{M}} \varphi^\circ$.

This builds a bridge between $\vDash_\rhd$ and $\vDash_>$. As an example:

Lemma 2 (translation). *Let M be the class of minimal models and $\circ = \circ_\gtrdot$. Then $M \overset{\circ}{\approx} M$.*

Proof. By induction on the complexity of the formula. The case $\alpha = p$ follows by definition. The cases for $\neg, \wedge, \vee, \rightarrow$ work by using the induction hypothesis (IH). Thus it suffices to prove the case for $\gtrdot$:

$$
\begin{array}{llll}
w \vDash_\gtrdot \varphi \gtrdot \psi & \text{iff} & [\psi]_\gtrdot \in F(w, [\varphi]_\gtrdot) \;\&\; [\bot]_\gtrdot \notin F(w, [\varphi]_\gtrdot) & (\gtrdot)\\
& \text{iff} & [\psi^\circ]_> \in F(w, [\varphi^\circ]_>) \;\&\; [\bot]_> \notin F(w, [\varphi^\circ]_>) & (\text{IH})\\
& \text{iff} & w \vDash_> (\varphi^\circ > \psi^\circ) \wedge \neg(\varphi^\circ > \bot) & (>)\\
& \text{iff} & w \vDash_> (\varphi \gtrdot \psi)^\circ & (\circ)
\end{array}
$$

□

Thus the translation $\circ$ is well behaved: The translate α° expresses the same proposition in $\mathcal{L}_>$ as the original formula α in $\mathcal{L}_\gtrdot$.

For the method to work, it will be essential to have a *backtranslation* $\bullet$ of $>$ into $\rhd$. For our example, I conjecture (and later show) that we can work with the following backtranslation:

($\bullet_\gtrdot$) $(\varphi > \psi)^\bullet := (\varphi^\bullet \gtrdot \psi^\bullet) \vee \neg(\varphi^\bullet \gtrdot \top)$

The backtranslation should also be semantically well behaved. For the neutral backtranslation to be well behaved, we have to consider minimal rcm-cn models with complete logic $\mathsf{CE} + \mathsf{Cm} + \mathsf{CN}$ (Theorem 3 and previous remarks).

Lemma 3 (backtranslation). *Let M be the class of minimal rcm-cn models and $\bullet = \bullet_\gtrdot$. Then $M \overset{\bullet}{\approx} M$.*

Proof. It suffices to prove the case for $>$:

$$
\begin{array}{llll}
w \vDash_> \varphi > \psi & \text{iff} & w \vDash_> ((\varphi > \psi) \wedge \neg(\varphi > \bot)) \vee ((\varphi > \psi) \wedge (\varphi > \bot)) & \\
& \text{iff} & w \vDash_> ((\varphi > \psi) \wedge \neg(\varphi > \bot)) \vee (\varphi > \bot) & (\mathsf{RCM})\\
& \text{iff} & w \vDash_> ((\varphi > \psi) \wedge \neg(\varphi > \bot)) \vee \neg(\varphi > \top) \vee (\varphi > \bot) & (\mathsf{CN})\\
& \text{iff} & w \vDash_> ((\varphi > \psi) \wedge \neg(\varphi > \bot)) \vee \neg((\varphi > \top) \wedge \neg(\varphi > \bot)) & \\
& \text{iff} & ([\psi]_> \in F(w, [\varphi]_>) \;\&\; [\bot]_> \notin F(w, [\varphi]_>)) & \\
& & \text{or not } ([\top]_> \in F(w, [\varphi]_>) \;\&\; [\bot]_> \notin F(w, [\varphi]_>)) & (>)\\
& \text{iff} & ([\psi^\bullet]_\gtrdot \in F(w, [\varphi^\bullet]_\gtrdot) \;\&\; [\bot]_\gtrdot \notin F(w, [\varphi^\bullet]_\gtrdot)) & \\
& & \text{or not } ([\top]_\gtrdot \in F(w, [\varphi^\bullet]_\gtrdot) \;\&\; [\bot]_\gtrdot \notin F(w, [\varphi^\bullet]_\gtrdot)) & (\text{IH})\\
& \text{iff} & w \vDash_\gtrdot (\varphi^\bullet \gtrdot \psi^\bullet) \vee \neg(\varphi^\bullet \gtrdot \top) & (\gtrdot)\\
& \text{iff} & w \vDash_\gtrdot (\varphi > \psi)^\bullet & (\bullet)
\end{array}
$$

□

The following section exploits the fact that the logic for $\rhd$ and that for $>$ is essentially the same, modulo translation:

Lemma 4. *Let* $\circ\colon \mathcal{L}_{\rhd} \longrightarrow \mathcal{L}_{>}$ *be a translation and* $M \overset{\circ}{\approx} M$. *Then* $M \vDash_{\rhd} \varphi$ *iff* $M \vDash_{>} \varphi^{\circ}$.

A proof of a more general result, can be found in Raidl (2020, Lemma 4.3).

4 Transferring

This section develops an indirect method to derive soundness, completeness and correspondence results of a defined conditional $\rhd$ from known soundness, completeness and correspondence results of the defining basic conditional $>$, using the translations between the two languages.[7]

Definition 5. Let $\circ\colon \mathcal{L}_{\rhd} \longrightarrow \mathcal{L}_{>}$ be a translation, and $\Gamma_{>}, \Gamma_{\rhd}$ axiomatic systems in $\mathcal{L}_{>}$ and $\mathcal{L}_{\rhd}$. $\Gamma_{>}$ *simulates* $\Gamma_{\rhd}$ *modulo* $\circ$, $\Gamma_{\rhd} \overset{\circ}{\propto} \Gamma_{>}$, iff for every $\alpha \in \mathcal{L}_{\rhd}$, $\vdash_{\Gamma_{\rhd}} \alpha$ implies $\vdash_{\Gamma_{>}} \alpha^{\circ}$.

This is a syntactic analogue to the semantic bridge $\overset{\circ}{\approx}$. Two things should be noted. First, for $\Gamma_{>}$ to *simulate* $\Gamma_{\rhd}$ *modulo* $\circ$, it suffices that all rules of $\Gamma_{\rhd}$ can be simulated by $\Gamma_{>}$ and all axioms of $\Gamma_{\rhd}$ can be simulated by $\Gamma_{>}$. We will use this obvious fact in what follows. Second, we have:

Lemma 5. *Let* $\Gamma_{>}, \Gamma_{\rhd}$ *be classical conditional logics and* $\circ : \mathcal{L}_{\rhd} \longrightarrow \mathcal{L}_{>}$ *a translation. Then the rules* MoPo, RCEA *and* RCEC *and the axiom scheme* PT *of* $\Gamma_{\rhd}$ *can be simulated in* $\Gamma_{>}$ *modulo* $\circ$.

Proof. By induction on the complexity of the formula.

PT: Let $\gamma \in \mathcal{L}_{\rhd}$ be a substitution instance of a classical tautology. Thus there is $\theta \in \mathcal{L}$ (a classical formula), $p_1, \ldots, p_n \in \text{Var}$, and $\varphi_1, \ldots, \varphi_n \in \mathcal{L}_{\rhd}$, such that $\gamma = \theta[\varphi_1/p_1, \ldots, \varphi_n/p_n]$. Therefore $\gamma^{\circ} = \theta[\varphi_1^{\circ}/p_1, \ldots, \varphi_n^{\circ}/p_n]$ (Lemma 1). This is a formula in $\mathcal{L}_{>}$ and a substitution instance of a classical tautology. Thus it is derivable in $\Gamma_{>}$, since PT is an axiom.

MoPo: The $\circ$-translation is MoPo$^{\circ}$: $\varphi^{\circ}, (\varphi \to \psi)^{\circ} \vdash \psi^{\circ}$. We need to prove this in $\Gamma_{>}$. Thus suppose φ° and $(\varphi \to \psi)^{\circ}$. But $(\varphi \to \psi)^{\circ} = (\varphi^{\circ} \to \psi^{\circ})$. Thus ψ° is provable in $\Gamma_{>}$ (MoPo is a rule).

RCEC: let $\alpha = \alpha[p, q]$ such that $(\varphi \rhd \psi)^{\circ} = \alpha[\varphi^{\circ}/p, \psi^{\circ}/q]$. The $\circ$-translation of RCEC is RCEC$^{\circ}$: If $\vdash (\psi \leftrightarrow \varphi)^{\circ}$ then $\vdash \alpha[\chi^{\circ}/p, \psi^{\circ}/q] \to \alpha[\chi^{\circ}/p, \varphi^{\circ}/q]$. But $(\psi \leftrightarrow \varphi)^{\circ} = (\psi^{\circ} \leftrightarrow \varphi^{\circ})$. Thus RCEC$^{\circ}$ is an instance of substitution of provable equivalents. To see this, consider $\alpha' = \alpha[\chi^{\circ}/p, \psi^{\circ}/q]$, then $\alpha[\chi^{\circ}/p, \varphi^{\circ}/q] = \alpha'[\varphi^{\circ}/\psi^{\circ}]$. But $\Gamma_{>}$ contains CE, thus substitution of provable equivalents holds. Therefore RCEC$^{\circ}$ holds in $\Gamma_{>}$.

RCEA can be simulated in a similar fashion as RCEC. □

[7] A more general account is developed in Raidl (2020).

The previously mentioned semantic and axiomatic bridges allow transferring soundness and completeness from $>$ to $\triangleright$:

Theorem 4 (Completeness Transfer). *Let M a model class and $\Gamma_{\triangleright}, \Gamma_{>}$ axiomatic systems in $\mathcal{L}_{\triangleright}, \mathcal{L}_{>}$ respectively (containing* MoPo*). Assume*

1. *$\Gamma_{>}$ is sound and complete for M in $\mathcal{L}_{>}$,*
2. *$\circ : \mathcal{L}_{\triangleright} \longrightarrow \mathcal{L}_{>}$ and $\bullet : \mathcal{L}_{>} \longrightarrow \mathcal{L}_{\triangleright}$ are translations,*
3. *$M \overset{\circ}{\approx} M$,* (Translation Lemma)
4. *$\Gamma_{>} \overset{\bullet}{\propto} \Gamma_{\triangleright}$ and $\Gamma_{\triangleright} \overset{\circ}{\propto} \Gamma_{>}$,* (Simulation Lemma)
5. *$\vdash_{\Gamma_{\triangleright}} \alpha^{\circ\bullet} \leftrightarrow \alpha$.* (Twin Lemma)

Then $\Gamma_{\triangleright}$ is sound and complete for M in $\mathcal{L}_{\triangleright}$.

Proof. Soundness Transfer: Suppose $\vdash_{\Gamma_{\triangleright}} \alpha$. Thus $\vdash_{\Gamma_{>}} \alpha^{\circ}$ (4). Hence $M \vDash_{>} \alpha^{\circ}$ (1). Therefore $M \vDash_{\triangleright} \alpha$ (3). *Completeness Transfer*: Suppose $M \vDash_{\triangleright} \alpha$. Then $M \vDash_{>} \alpha^{\circ}$ (3). Thus $\vdash_{\Gamma_{>}} \alpha^{\circ}$ (1). Therefore $\vdash_{\Gamma_{\triangleright}} \alpha^{\circ\bullet}$ (4). Hence $\vdash_{\Gamma_{\triangleright}} \alpha$ (5, MoPo). □

An application is essentially given by a particular pair of translations $\circ$ and $\bullet$ and consists in proving (3), (4) and (5). The real work however is to figure out a well-behaved backtranslation $\bullet$, from which one can then roughly obtain $\Gamma_{\gtrdot}$ by backtranslating all axioms of $\Gamma_{>}$.[8]

Let me illustrate the Twin Lemma (5) in the above Theorem with the neutral conditional, which roughly says that χ is equivalent to its *twin* $\chi^{\circ\bullet}$ in the conjectured logic for $\gtrdot$:

Lemma 6 (Twin). *$\vdash_{\mathsf{CM+Con}} \chi^{\circ\bullet} \leftrightarrow \chi$, where $\circ = \circ_{\gtrdot}$ and $\bullet = \bullet_{\gtrdot}$.*

Proof. By induction on the complexity of the formula. It suffices to verify the case for $\gtrdot$. Recall, CM is the smallest consequent monotone conditional logic, and Con the axiom $\neg(\varphi \gtrdot \bot)$. Let $\equiv$ denote provable equivalence in CM + Con:

$$
\begin{aligned}
& ((\varphi \gtrdot \psi)^{\circ})^{\bullet} \\
&= (\varphi^{\circ} > \psi^{\circ})^{\bullet} \wedge \neg(\varphi^{\circ} > \bot)^{\bullet} && (\circ, \bullet) \\
&\equiv (\neg(\varphi \gtrdot \top) \vee (\varphi \gtrdot \psi)) \wedge \neg(\neg(\varphi \gtrdot \top) \vee (\varphi \gtrdot \bot)) && (\bullet, \mathrm{IH}, \mathsf{RCEA}, \mathsf{RCEC}) \\
&\equiv (\neg(\varphi \gtrdot \top) \vee (\varphi \gtrdot \psi)) \wedge (\varphi \gtrdot \top) \wedge \neg(\varphi \gtrdot \bot) \\
&\equiv (\neg(\varphi \gtrdot \top) \vee (\varphi \gtrdot \psi)) \wedge (\varphi \gtrdot \top) && (\mathsf{Con}) \\
&\equiv (\varphi \gtrdot \psi) \wedge (\varphi \gtrdot \top) \\
&\equiv (\varphi \gtrdot \psi) && (\mathsf{RCM})
\end{aligned}
$$

□

Similarly, we can illustrate the Simulation Lemma (4) from Theorem 4:

Lemma 7 (Simulation). *Let $\circ = \circ_{\gtrdot}$ and $\bullet = \bullet_{\gtrdot}$. Then*

1. CM + Con $\overset{\circ}{\propto}$ CM + CN.
2. CM + CN $\overset{\bullet}{\propto}$ CM + Con.

[8] Compare the 'fixed-point heuristics' in Raidl (2020).

Proof. (1) We have $\mathsf{CM} = \mathsf{CE} + \mathsf{RCM}$. Simulation of CE follows from Lemma 5. The simulation of RCM can be shown as we did for RCEC. Additionally, Con° is the tautology $\neg(\varphi > \bot) \vee (\varphi > \bot)$. (2) Simulation of CM again by Lemma 5. $\mathsf{CN}^{\bullet}$ is the tautology $(\varphi \gg \top) \vee \neg(\varphi \gg \top)$. □

The transfer of the correspondence theory is more complicated. Let X be an axiom scheme in $\mathcal{L}_{>}$. We write $w \vDash \mathsf{X}$ iff $w \vDash \varphi$ for all $\varphi \in \mathsf{X}$. This lifts truth from formulas to axiom schemes. We write $M \vDash \mathsf{X}$ iff all models in M validate all instances of X. Finally, we write $M \vDash \mathsf{X} \equiv \mathsf{X}'$ iff for all $\mathfrak{M} \in M$ and all $w \in W$, we have $w \vDash^{\mathfrak{M}} \mathsf{X}$ iff $w \vDash^{\mathfrak{M}} \mathsf{X}'$.

To transform an axiom scheme X holding for $>$ to an axiom scheme $\mathsf{X}_{\rhd}$ holding for $\rhd$ we first backtranslate X into $\mathsf{X}^{\bullet} = \{\varphi^{\bullet} : \varphi \in \mathsf{X}\}$, and second clean this backtranslate to obtain an equivalent nicer looking axiom scheme $\mathsf{X}_{\rhd}$.

Theorem 5 (Axiom transfer). *With M as previously and $\mathsf{X}_{\rhd}$, X axiom schemes in $\mathcal{L}_{\rhd}$, $\mathcal{L}_{>}$ respectively. Assume that*

1. $\bullet : \mathcal{L}_{>} \longrightarrow \mathcal{L}_{\rhd}$ *is a translation,*
2. $M \overset{\bullet}{\approx} M$, (Backtranslation Lemma)
3. $M \vDash \mathsf{X}_{\rhd} \equiv \mathsf{X}^{\bullet}$. (Cleaning)

Then for $N \subseteq M$, $N \vDash_{>} \mathsf{X}$ iff $N \vDash_{\rhd} \mathsf{X}_{\rhd}$.

Proof. First note that $M \overset{\bullet}{\approx} M$ and $N \subseteq M$ implies $N \overset{\bullet}{\approx} N$. Thus, we have:

$$\begin{aligned} N \vDash_{\rhd} \mathsf{X}_{\rhd} \ &\text{iff}\ \ N \vDash_{\rhd} \mathsf{X}^{\bullet}\ (M \vDash_{\rhd} \mathsf{X}_{\rhd} \equiv \mathsf{X}^{\bullet} \text{ and } N \subseteq M) \\ &\text{iff}\ \ N \vDash_{>} \mathsf{X} \qquad\qquad (N \overset{\bullet}{\approx} N, \text{Lemma 4}) \end{aligned}$$

□

The results presented here allow to generate a sound and complete logic $\Gamma_{\rhd}$ for a defined conditional $\rhd$, based on the sound and complete logic $\Gamma_{>}$ for the defining conditional $>$. The last theorem also allows to transfer the correspondence theory from $>$ to $\rhd$.

5 Neutral Conditional

This section applies the results of the two previous sections to the neutral conditional ($\mathcal{L}_{\rhd} = \mathcal{L}_{\gg}$). It illustrates how to establish the premisses of the two theorems from the previous section in a particular case. Other examples can be developed along similar lines.

Let me briefly motivate the neutral conditional. *Vacuism* is the position that all conditionals with impossible antecedent are true (Williamson 2007). That is, vacuist logics validate

$$\neg \Diamond \varphi \rightarrow (\varphi > \psi)$$

Neutralism treats all impossible antecedent conditionals as false but maintains RCM. As a consequence, neutralism needs to reject all conditionals $\varphi > \bot$. This means imposing:

$$\neg(\varphi > \bot) \qquad \text{Con}$$

The neutral conditional validates Con. The neutral conditional was coined by Lewis (1973b, 24-6), and investigated in the possibilistic and ranking semantics (Benferhat, Dubois, & Prade, 1997; Dubois & Prade, 1994; Raidl, 2019a).

Here is our proposal for a minimal logic for the neutral conditional:

Definition 6. $\mathsf{E} := \mathsf{CM} + \mathsf{Con}$ is the smallest neutral conditional logic.[9]

Using the results of Section 3 and 4, we can prove:

Theorem 6. E *is sound and complete for* $>$ *in the minimal rcm-cn models.*

Proof. We establish (1)–(5) of Theorem 4. Let M be the minimal rcm-cn models. Recall $\mathsf{CM} = \mathsf{CE} + \mathsf{RCM}$. (1) $\mathsf{CM} + \mathsf{CN}$ is sound and complete for M in $\mathcal{L}_>$ by Theorem 3. (2) holds trivially. (3) $M \overset{\circ}{\approx} M$ by the Translation Lemma 2. (4) by the Simulation Lemma 7 and (5) $\mathsf{CM} + \mathsf{Con} \vdash \chi^{\circ\bullet} \leftrightarrow \chi$ by the Twin Lemma 6. Completeness and soundness follow from Theorem 4. □

The neutral conditional logic E and its vacuist source logic CM + CN are both consequent-monotone conditional logics. They only differ in the law of conditional consistency Con and the law of trivial conditionals CN. As one can strengthen the vacuist logic CM + CN, one can strengthen its neutral analogue E. For this, we will use Theorem 5.

We use the following procedure: Given a source scheme X for $>$ we backtranslate it first into $\mathsf{X}^\bullet = \{\varphi^\bullet : \varphi \in \mathsf{X}\}$. We then obtain our proposed neutral analogue X^E by cleaning $\mathsf{X}^\bullet$, i.e., applying logical transformations in E to simplify the backtranslate which often looks quite ugly. This procedure – backtranslating + cleaning – may end in a conjunction of axiom schemes, which we then denote $\mathsf{X1}^\mathsf{E}, \mathsf{X2}^\mathsf{E}, \ldots$.

Theorem 7. *Let* M *be the rcm-cn minimal models and* $N \subseteq M$. *Then* $N \vDash_> \mathsf{X}$ *iff* $N \vDash_> \mathsf{X}^\mathsf{E}$, *where* X *is one of the axiom schemes on the left of the following table and* X^E *is the corresponding (conjunction of) scheme(s) on the right:*

[9] Recall $\mathsf{CM} = \mathsf{CE} + \mathsf{RCM} = \mathsf{CE} + \mathsf{Cm}$ (see Section 2).

X		X^E
CC	CC	CC^E
M	$(\varphi > \top) \to ((\varphi \vee \psi) > \top)$	M^E
C	$((\varphi \vee \psi) > \top) \to ((\varphi > \top) \vee (\psi > \top))$	C^E
ID	$(\varphi > \top) \to (\varphi > \varphi)$	ID^E
CV	$((\varphi \wedge \psi) > \top) \to (((\varphi > \chi) \wedge \neg(\varphi > \neg\psi)) \to ((\varphi \wedge \psi) > \chi))$	CV^E
CMon	$((\varphi \wedge \psi) > \top) \to (((\varphi > \chi) \wedge (\varphi > \psi)) \to ((\varphi \wedge \psi) > \chi))$	$\mathsf{CMon1}^\mathsf{E}$
	M^E	$\mathsf{CMon2}^\mathsf{E}$
CA	$((\varphi \vee \psi) > \top) \to (((\varphi > \chi) \wedge (\psi > \chi)) \to ((\varphi \vee \psi) > \chi))$	$\mathsf{CA1}^\mathsf{E}$
	$((\varphi \vee \psi) > \top) \to (((\varphi > \chi) \wedge \neg(\psi > \top)) \to ((\varphi \vee \psi) > \chi))$	$\mathsf{CA2}^\mathsf{E}$
	C^E	$\mathsf{CA3}^\mathsf{E}$
WCut	$(\varphi > \psi) \to ((\varphi \wedge \psi) > \top)$	WCut^E
Cut	$\mathsf{Cut} + \mathsf{WCut}^\mathsf{E}$	Cut^E

The theorem says that a class N of rcm-cn minimal models validates the axiom scheme X with respect to the standard truth clause for $>$ iff the same class validates our neutral analogue X^E with respect to the truth clause for $>$. The proof illustrates how to simplify the backtranslate $\mathsf{X}^\bullet$ into a nicer looking X^E.

Proof. We establish (1)–(3) of Theorem 5. (1) is clear. (2) $M \overset{\bullet}{\approx} M$ by Lemma 3. Thus it remains to prove (3), i.e., that X^E is scheme-equivalent to $\mathsf{X}^\bullet$, in $\mathsf{E} = \mathsf{CM} + \mathsf{Con}$. Scheme equivalence in E is denoted by $\equiv$.

$$\begin{aligned}\mathsf{CC}^\bullet &= (((\varphi > \psi) \vee \neg(\varphi > \top)) \wedge ((\varphi > \chi) \vee \neg(\varphi > \top))) \to ((\varphi > (\psi \wedge \chi)) \vee \neg(\varphi > \top))\\ &\equiv ((\varphi > \psi) \wedge (\varphi > \chi) \wedge (\varphi > \top)) \to (\varphi > (\psi \wedge \chi))\\ &\equiv ((\varphi > \psi) \wedge (\varphi > \chi)) \to (\varphi > (\psi \wedge \chi)) && (\mathsf{RCM})\\ &= \mathsf{CC} = \mathsf{CC}^\mathsf{E}\end{aligned}$$

$$\mathsf{ID}^\bullet = ((\varphi > \varphi) \vee \neg(\varphi > \top)) \equiv ((\varphi > \top) \to (\varphi > \varphi)) = \mathsf{ID}^\mathsf{E}.$$

$$\begin{aligned}\mathsf{M}^\bullet &= ((\alpha > \bot) \vee \neg(\alpha > \top)) \to (((\alpha \wedge \beta) > \bot) \vee \neg((\alpha \wedge \beta) > \top))\\ &\equiv \neg(\alpha > \top) \to \neg((\alpha \wedge \beta) > \top) && (\mathsf{Con})\\ &\equiv ((\alpha \wedge \beta) > \top) \to (\alpha > \top)\\ &\equiv (\varphi > \top) \to ((\varphi \vee \psi) > \top) && (\mathsf{CE})\\ &= \mathsf{M}^\mathsf{E}\end{aligned}$$

$$\begin{aligned}\mathsf{C}^\bullet &= (((\varphi > \bot) \vee \neg(\varphi > \top)) \wedge ((\psi > \bot) \vee \neg(\psi > \top)))\\ &\quad \to (((\varphi \vee \psi) > \bot) \vee \neg((\varphi \vee \psi) > \top))\\ &\equiv (\neg(\varphi > \top) \wedge \neg(\psi > \top)) \to \neg((\varphi \vee \psi) > \top) && (\mathsf{Con})\\ &\equiv ((\varphi \vee \psi) > \top) \to ((\varphi > \top) \vee (\psi > \top))\\ &= \mathsf{C}^\mathsf{E}\end{aligned}$$

$\mathsf{CV}^\bullet$ is equivalent to

$$\left(\begin{pmatrix} \varphi > \chi \\ \vee\ \neg(\varphi > \top)\end{pmatrix} \wedge \begin{pmatrix} \neg(\varphi > \neg\psi) \\ \wedge\quad \varphi > \top\end{pmatrix}\right) \to \begin{pmatrix} (\varphi \wedge \psi) > \chi \\ \vee\ \neg((\varphi \wedge \psi) > \top)\end{pmatrix}$$

$\neg(\varphi > \top)$ contradicts $\varphi > \top$. $\varphi > \chi$ implies $\varphi > \top$ (RCM). Thus,

$$\begin{aligned}\mathsf{CV}^\bullet &\equiv ((\varphi > \chi) \wedge \neg(\varphi > \neg\psi)) \to (((\varphi \wedge \psi) > \chi) \vee \neg((\varphi \wedge \psi) > \top))\\ &\equiv ((\varphi > \chi) \wedge \neg(\varphi > \neg\psi) \wedge ((\varphi \wedge \psi) > \top)) \to ((\varphi \wedge \psi) > \chi)\\ &\equiv ((\varphi \wedge \psi) > \top) \to (((\varphi > \chi) \wedge \neg(\varphi > \neg\psi)) \to ((\varphi \wedge \psi) > \chi))\\ &= \mathsf{CV}^\mathsf{E}\end{aligned}$$

$\mathsf{CMon}^\bullet$ is

$$\left(\left(\begin{array}{c}\varphi > \chi \\ \vee \neg(\varphi > \top)\end{array}\right) \wedge \left(\begin{array}{c}\varphi > \psi \\ \vee \neg(\varphi > \top)\end{array}\right)\right) \rightarrow \left(\begin{array}{c}((\varphi \wedge \psi) > \chi) \\ \vee \neg((\varphi \wedge \psi) > \top)\end{array}\right)$$

$(\varphi > \chi) \wedge \neg(\varphi > \top)$ as $(\varphi > \psi) \wedge \neg(\varphi > \top)$ contradict RCM. Hence $\mathsf{CMon}^\bullet$ is equivalent to

$$\left(\left(\begin{array}{c}(\varphi > \chi) \wedge (\varphi > \psi) \\ \vee \quad \neg(\varphi > \top)\end{array}\right) \wedge ((\varphi \wedge \psi) > \top)\right) \rightarrow ((\varphi \wedge \psi) > \chi)$$

This is equivalent to the conjunction of

1. $((\varphi > \chi) \wedge (\varphi > \psi) \wedge ((\varphi \wedge \psi) > \top)) \rightarrow ((\varphi \wedge \psi) > \chi)$
2. $(\neg(\varphi > \top) \wedge ((\varphi \wedge \psi) > \top)) \rightarrow ((\varphi \wedge \psi) > \chi)$

(1) is equivalent to $\mathsf{CMon1^E}$. (2) is equivalent to:

$$\begin{array}{ll}\equiv (\neg(\varphi > \top) \wedge ((\varphi \wedge \psi) > \top)) \rightarrow ((\varphi \wedge \psi) > \bot) & (\mathsf{RCM}) \\ \equiv \neg(\neg(\varphi > \top) \wedge ((\varphi \wedge \psi) > \top)) & (\mathsf{Con}) \\ \equiv ((\varphi \wedge \psi) > \top) \rightarrow (\varphi > \top) & \\ \equiv (\alpha > \top) \rightarrow ((\alpha \vee \beta) > \top) & (\mathsf{CE}) \\ = \mathsf{CMon2^E} = \mathsf{M^E} & \end{array}$$

$\mathsf{CA}^\bullet$ is equivalent (similar reasoning as for $\mathsf{CMon}^\bullet$) to the conjunction of:

1. $((\varphi > \chi) \wedge (\psi > \chi) \wedge ((\varphi \vee \psi) > \top)) \rightarrow ((\varphi \vee \psi) > \chi)$
2. $((\varphi > \chi) \wedge \neg(\psi > \top) \wedge ((\varphi \vee \psi) > \top)) \rightarrow ((\varphi \vee \psi) > \chi)$
3. $(\neg(\varphi > \top) \wedge (\psi > \chi) \wedge ((\varphi \vee \psi) > \top)) \rightarrow ((\varphi \vee \psi) > \chi)$
4. $(\neg(\varphi > \top) \wedge \neg(\psi > \top) \wedge ((\varphi \vee \psi) > \top)) \rightarrow ((\varphi \vee \psi) > \chi)$

(1) is equivalent to $\mathsf{CA1^E}$. (2) is equivalent to $\mathsf{CA2^E}$. (3) has the same form. (4) is equivalent to the following (similar reasoning as for $\mathsf{CMon2^E}$):

$$\begin{array}{ll}\equiv (\neg(\varphi > \top) \wedge \neg(\psi > \top) \wedge ((\varphi \vee \psi) > \top)) \rightarrow ((\varphi \vee \psi) > \bot) & (\mathsf{RCM}) \\ \equiv \neg(\neg(\varphi > \top) \wedge \neg(\psi > \top) \wedge ((\varphi \vee \psi) > \top)) & (\mathsf{Con}) \\ \equiv ((\varphi \vee \psi) > \top) \rightarrow ((\varphi > \top) \vee (\psi > \top)) & \\ = \mathsf{CA3^E} = \mathsf{C^E} & \end{array}$$

$\mathsf{WCut}^\bullet$ is equivalent to

$$\left(\begin{array}{l}((\varphi > \psi) \vee \neg(\varphi > \top)) \\ \wedge (((\varphi \wedge \psi) > \bot) \vee \neg((\varphi \wedge \psi) > \top))\end{array}\right) \rightarrow ((\varphi > \bot) \vee \neg(\varphi > \top))$$

$\varphi > \bot$ and $(\varphi \wedge \psi) > \bot$ contradict Con. And $\neg(\varphi > \top)$ in the antecedent implies trivially $\neg(\varphi > \top)$ in the consequent. Thus $\mathsf{WCut}^\bullet$ is equivalent to:

$$\begin{array}{ll}((\varphi > \psi) \wedge \neg((\varphi \wedge \psi) > \top)) \rightarrow \neg(\varphi > \top) & \\ \equiv ((\varphi > \psi) \wedge (\varphi > \top)) \rightarrow ((\varphi \wedge \psi) > \top) & \\ \equiv (\varphi > \psi) \rightarrow ((\varphi \wedge \psi) > \top) & (\mathsf{RCM}) \\ = \mathsf{WCut^E} & \end{array}$$

$\mathsf{Cut}^\bullet$ is equivalent to

$$\left(\left(\begin{array}{c}((\varphi \gtrdot \psi) \vee \neg(\varphi \gtrdot \top)) \\ \wedge\ (((\varphi \wedge \psi) \gtrdot \chi) \vee \neg((\varphi \wedge \psi) \gtrdot \top))\end{array}\right) \wedge (\varphi \gtrdot \top)\right) \rightarrow (\varphi \gtrdot \chi)$$

$\neg(\varphi \gtrdot \top)$ contradicts $\varphi \gtrdot \top$. And $\varphi \gtrdot \psi$ implies $\varphi \gtrdot \top$ (RCM). Thus $\mathsf{Cut}^\bullet$ is equivalent to the conjunction of:

1. $((\varphi \gtrdot \psi) \wedge ((\varphi \wedge \psi) \gtrdot \chi)) \rightarrow (\varphi \gtrdot \chi)$
2. $((\varphi \gtrdot \psi) \wedge \neg((\varphi \wedge \psi) \gtrdot \top)) \rightarrow (\varphi \gtrdot \chi)$

(1) is Cut. (2) is equivalent to the following (similar reasoning as for $\mathsf{CMon}^\bullet$):

$$\begin{array}{ll}\equiv ((\varphi \gtrdot \psi) \wedge \neg((\varphi \wedge \psi) \gtrdot \top)) \rightarrow (\varphi \gtrdot \bot) & (\mathsf{RCM}) \\ \equiv \neg((\varphi \gtrdot \psi) \wedge \neg((\varphi \wedge \psi) \gtrdot \top)) & (\mathsf{Con}) \\ \equiv (\varphi \gtrdot \psi) \rightarrow ((\varphi \wedge \psi) \gtrdot \top) & \\ = \mathsf{WCut}^\mathsf{E} & \end{array}$$

□

Let me explain the new axioms. The only axiom which remains unchanged under the translation is the axiom CC (modulo provable equivalence in E). That is $\mathsf{CC}^\bullet \equiv \mathsf{CC}(\gtrdot)$ in E, where for clarity $\mathsf{CC}(\gtrdot)$ is the axiom scheme CC formulated with $\gtrdot$ instead of $>$.

To explain the remaining axioms, we need to analyse the outer modalities in $\mathcal{L}_\gtrdot$, which are now expressed by $\Diamond^\mathsf{E}\alpha := \alpha \gtrdot \top$ and $\Box^\mathsf{E}\alpha := \neg(\neg\alpha \gtrdot \top)$. This can be seen as follows. In $\mathcal{L}_>$ the outer possibility of α° is $\neg(\alpha^\circ > \bot)$. Consider the backtranslate $(\neg(\alpha^\circ > \bot))^\bullet$. Resolving the backtranslation, we obtain first $\neg(\alpha^\circ > \bot)^\bullet$, then $\neg((\alpha^{\circ\bullet} \gtrdot \bot) \vee \neg(\alpha^{\circ\bullet} \gtrdot \top))$, which resolves into $\neg(\alpha^{\circ\bullet} \gtrdot \bot) \wedge (\alpha^{\circ\bullet} \gtrdot \top)$ and can be reduced to $\alpha^{\circ\bullet} \gtrdot \top$ by Con. But α is equivalent to its twin $\alpha^{\circ\bullet}$ in E, by the Twin Lemma 6. Thus $\alpha^{\circ\bullet} \gtrdot \top$ reduces to $\alpha \gtrdot \top$, which is the new outer possibility $\Diamond^\mathsf{E}\alpha$. Dually $\Box^\mathsf{E}\alpha = \neg(\neg\alpha \gtrdot \top)$.

It is now evident, that if we reformulate M^E with respect to $\Box^\mathsf{E}$, it will have the same form as the outer necessity reformulation M' of M (p. 3). The same can be said for C^E. ID^E says that if φ is possible then $\varphi \gtrdot \varphi$ holds. The new outer possibility can then equivalently be expressed by $\varphi \gtrdot \varphi$ instead of $\varphi \gtrdot \top$. A similar remark holds for CV^E, which says that if $\varphi \wedge \psi$ is possible then an instance of the axiom CV holds for $\gtrdot$.

To make this precise, let us describe it by a general transformation. For X a scheme of the form $A \rightarrow (B > C)$, we denote $\Diamond\mathsf{X}$ the scheme prefixed by the possibility of the antecedent of the consequent conditional of X, that is $\Diamond B \rightarrow (A \rightarrow (B > C))$. Call this operation *prefixing*. The axioms ID^E and CV^E are obtained by prefixing. They have the form $\Diamond$ID and $\Diamond$CV.[10] Similarly $\mathsf{CMon1}^\mathsf{E}$ and $\mathsf{CA1}^\mathsf{E}$ are of the form $\Diamond$CMon and $\Diamond$CA. Thus transferring the axioms ID, CV, CMon and CA to $\mathcal{L}_\gtrdot$ yields a prefixed version of the original (partly for the latter two).

[10] To be entirely precise, they have the form $\Diamond^\mathsf{E}\mathsf{ID}(\gtrdot)$ and $\Diamond^\mathsf{E}\mathsf{CV}(\gtrdot)$.

In $\mathsf{CMon}^{\mathsf{E}}$ a second axiom $\mathsf{CMon2}^{\mathsf{E}} = \mathsf{M}^{\mathsf{E}}$ pops up. Thus, given M^{E}, $\mathsf{CMon}^{\mathsf{E}}$ is equivalent to just $\Diamond\mathsf{CMon}$. Furthermore, in presence of $\mathsf{WCut}^{\mathsf{E}}$, $\Diamond\mathsf{CMon}$ reduces obviously to CMon. Similar remarks hold for CA^{E}. First, $\mathsf{CA1}^{\mathsf{E}} = \Diamond\mathsf{CA}$ and $\mathsf{CA3}^{\mathsf{E}} = \mathsf{C}^{\mathsf{E}}$. Second, given M^{E} (and RCM), $\Diamond\mathsf{CA}$ reduces to CA and $\mathsf{CA2}^{\mathsf{E}}$ to

$$\mathsf{CAI}\quad ((\varphi > \chi) \wedge \neg(\psi > \top)) \rightarrow ((\varphi \vee \psi) > \chi).$$

Reformulated with the outer modality, this is just CA where $\psi > \chi$ is replaced by the assumption that ψ is impossible – hence the 'I' in CAI.

In short, we have just proven the first three claims of:

Lemma 8. *In the system* E*, we have:*

1. $\mathsf{M}^{\mathsf{E}} \vdash \mathsf{CMon}^{\mathsf{E}} \equiv \Diamond\mathsf{CMon}$.
2. $\mathsf{WCut}^{\mathsf{E}} \vdash \Diamond\mathsf{CMon} \equiv \mathsf{CMon}$.
3. $\mathsf{M}^{\mathsf{E}} \vdash \mathsf{CA}^{\mathsf{E}} \equiv (\mathsf{CA} + \mathsf{CAI} + \mathsf{C}^{\mathsf{E}})$.
4. $\Diamond\mathsf{CV}, \mathsf{CC} \vdash \Diamond\mathsf{CMon}$.
5. $\mathsf{C}^{\mathsf{M}} + \Diamond\mathsf{ID} + \mathsf{CAI} + \mathsf{CC} \vdash \mathsf{WCut}^{\mathsf{E}}$.
6. $\mathsf{C}^{\mathsf{E}} + \Diamond\mathsf{ID} + \mathsf{CAI} \vdash \Diamond\mathsf{CV} \equiv \mathsf{CV}$.
7. $\mathsf{CV} + \mathsf{CC} \vdash \mathsf{C}^{\mathsf{E}}$.

Proof. It remains to prove (4)–(7).

(4) Assume CC and $\Diamond\mathsf{CV} = \mathsf{CV}^{\mathsf{E}}$. In $\mathsf{E} + \mathsf{CC}$, $\varphi > \psi$ implies $\neg(\varphi > \neg\psi)$ (by Con). Let's derive $\Diamond\mathsf{CMon}$. Suppose $(\varphi \wedge \psi) > \top$, $\varphi > \psi$ and $\varphi > \chi$, then we have $\neg(\varphi > \neg\psi)$ (previous remark). Hence $(\varphi \wedge \psi) > \chi$ ($\Diamond\mathsf{CV}$).

(5) Assume the mentioned axioms. Suppose $\varphi > \psi$ and (for reductio) $\neg((\varphi \wedge \psi) > \top)$. Since $\varphi > \psi$, we have $\varphi > \top$ (RCM). That is, $((\varphi \wedge \psi) \vee (\varphi \wedge \neg\psi)) > \top$ (RCEA). Thus $(\varphi \wedge \psi) > \top$ or $(\varphi \wedge \neg\psi) > \top$ (C^{E}). Therefore $(\varphi \wedge \neg\psi) > \top$ (by the reductio assumption). Hence $(\varphi \wedge \neg\psi) > (\varphi \wedge \neg\psi)$ ($\Diamond\mathsf{ID}$) and $\neg((\varphi \wedge \psi) > \top)$. Therefore $\varphi > (\varphi \wedge \neg\psi)$ ($\mathsf{CAI}, \mathsf{RCEA}$). But $\varphi > \psi$ (assumption) and thus $\varphi > \bot$ ($\mathsf{CC}, \mathsf{RCEC}$). This contradicts Con.

(6) Clearly CV implies $\Diamond\mathsf{CV}$. Conversely, let us first prove

$$\mathsf{CV0}\quad ((\varphi > \top) \wedge \neg(\varphi > \neg\psi)) \rightarrow ((\varphi \wedge \psi) > \top)$$

by proving a contraposed version: Assume $a = (\varphi > \top)$ and $b = \neg((\varphi \wedge \psi) > \top)$. We prove $\varphi > \neg\psi$. Either $i = \neg((\varphi \wedge \neg\psi) > \top)$ or $j = ((\varphi \wedge \neg\psi) > \top)$. If i, then since b, we obtain $\neg(\varphi > \top)$ (C^{E}). This contradicts a. Thus j. Hence $(\varphi \wedge \neg\psi) > \neg\psi$ ($\Diamond\mathsf{ID}$, RCM). Together with b, this implies $\varphi > \neg\psi$ (CAI, RCEA). This proves $\mathsf{CV0}$.

$\Diamond\mathsf{CV}$ implies CV: Assume $\varphi > \chi$ and $\neg(\varphi > \neg\psi)$. Thus $\varphi > \top$ (RCM), hence $(\varphi \wedge \psi) > \top$ ($\mathsf{CV0}$). And thus we can use $\Diamond\mathsf{CV}$ to obtain $(\varphi \wedge \psi) > \chi$.

(7) Suppose $(\varphi \vee \psi) > \top$. We cannot have both, $i = ((\varphi \vee \psi) > (\neg\varphi \wedge \psi))$ and $j = ((\varphi \vee \psi) > (\varphi \wedge \neg\psi))$. Else $(\varphi \vee \psi) > \bot$ (CC), contradicting Con. Thus $\neg i$ or $\neg j$. If $\neg i$, we obtain $\neg((\varphi \vee \psi) > \neg(\varphi \vee \neg\psi))$, but since $(\varphi \vee \psi) > \top$, we obtain $((\varphi \vee \psi) \wedge (\varphi \vee \neg\psi)) > \top$ (CV). Hence $\varphi > \top$ (RCEA, since $((\varphi \vee \psi) \wedge (\varphi \vee \neg\psi)) \equiv \varphi$). A similar reasoning from $\neg j$ yields $\psi > \top$. Thus $(\varphi > \top) \vee (\psi > \top)$. □

Using our axiom simplifications in extensions of E, we obtain neutral analogues for $\gtrdot$ to known conditional logics for $>$:

Corollary 1. *The neutral conditional $\gtrdot$ defined in the semantics on the left (with sound and complete logic for $>$ mentioned on the left) has the sound and complete logic on the right:*

semantics ($>$ logic)	$\gtrdot$ logic
standard (CK)	$\mathsf{E} + \mathsf{CC}$
cumulative (CU)	$\mathsf{E} + \diamondsuit\,\mathsf{ID} + \mathsf{Cut} + \mathsf{WCut}^{\mathsf{E}} + \mathsf{M}^{\mathsf{E}} + \mathsf{CMon}$
preferential (P)	$\mathsf{E} + \diamondsuit\,\mathsf{ID} + \mathsf{Cut} + \mathsf{Wcut}^{\mathsf{E}} + \mathsf{M}^{\mathsf{E}} + \mathsf{CMon} + \mathsf{C}^{\mathsf{E}} + \mathsf{CA} + \mathsf{CAI}$
Lewisean (V)	$\mathsf{E} + \diamondsuit\,\mathsf{ID} + \mathsf{CC} + \mathsf{M}^{\mathsf{E}} + \mathsf{CA} + \mathsf{CAI} + \mathsf{CV}$

Note that the axiomatic system for the neutral conditional obtained in Raidl (2019a, Theorem 5.1) in a semantics having the basic logic V augmented by the axiom $\neg(\top > \bot)$ has one redundant axiom, namely CMon, which is derivable from the remaining axioms, as seen in Lemma 8.

Proof. $\mathsf{CK} = \mathsf{CM} + \mathsf{CN} + \mathsf{CC}$. Denote M the standard models. Chellas (1975) proved that CK is sound and complete for M (in $\mathcal{L}_{>}$). By Theorem 7, we obtain that $\mathsf{E} + \mathsf{CC}^{\mathsf{E}}$ is sound and complete for M in $\mathcal{L}_{\gtrdot}$. And we know that this logic can be rewritten as $\mathsf{E} + \mathsf{CC}$.

It is known that $\mathsf{CU} = \mathsf{CM} + \mathsf{ID} + \mathsf{Cut} + \mathsf{CMon}$ is sound and complete for cumulative models (Kraus et al., 1990). By a similar reasoning as above, the corresponding logic for the neutral conditional is $\mathsf{E} + \diamondsuit\,\mathsf{ID} + \mathsf{Cut} + \mathsf{WCut}^{\mathsf{E}} + \mathsf{M}^{\mathsf{E}} + \diamondsuit\,\mathsf{CMon}$. Using Lemma 8.2, $\diamondsuit\,\mathsf{CMon}$ reduces to CMon, given $\mathsf{WCut}^{\mathsf{E}}$. Thus we obtain our logic $\mathsf{E} + \diamondsuit\,\mathsf{ID} + \mathsf{M}^{\mathsf{E}} + \mathsf{Cut} + \mathsf{WCut}^{\mathsf{E}} + \mathsf{CMon}$.

It is known that $\mathsf{P} = \mathsf{CU} + \mathsf{CA}$ is sound and complete for preferential models (Kraus et al., 1990). Using the previous result, the corresponding logic for the neutral conditional is $\mathsf{E} + \diamondsuit\,\mathsf{ID} + \mathsf{M}^{\mathsf{E}} + \mathsf{Cut} + \mathsf{WCut}^{\mathsf{E}} + \mathsf{CMon} + \mathsf{CA}^{\mathsf{E}}$. Using Lemma 8.3, and given M^{E}, CA^{E} reduces further to $\mathsf{CA} + \mathsf{CAI} + \mathsf{C}^{\mathsf{E}}$. Thus we get the logic $\mathsf{E} + \diamondsuit\,\mathsf{ID} + \mathsf{M}^{\mathsf{E}} + \mathsf{C}^{\mathsf{E}} + \mathsf{Cut} + \mathsf{WCut}^{\mathsf{E}} + \mathsf{CMon} + \mathsf{CA} + \mathsf{CAI}$.

It is known that $\mathsf{V} = \mathsf{CK} + \mathsf{ID} + \mathsf{CMon} + \mathsf{CA} + \mathsf{CV}$ is sound for (non-centered) Lewisean models (Lewis, 1971). By a similar reasoning as above, the corresponding logic for the neutral conditional is $\mathsf{E} + \mathsf{CC}^{\mathsf{E}} + \mathsf{ID}^{\mathsf{E}} + \mathsf{CMon}^{\mathsf{E}} + \mathsf{CA}^{\mathsf{E}} + \mathsf{CV}^{\mathsf{E}}$. But $\mathsf{CC}^{\mathsf{E}} = \mathsf{CC}$, $\mathsf{ID}^{\mathsf{E}} = \diamondsuit\,\mathsf{ID}$, $\mathsf{CMon}^{\mathsf{E}} = \mathsf{M}^{\mathsf{E}} + \diamondsuit\,\mathsf{CMon}$ and $\mathsf{CV}^{\mathsf{E}} = \diamondsuit\,\mathsf{CV}$. By (3) of Lemma 8, given M^{E}, CA^{E} is equivalent to $\mathsf{CA} + \mathsf{CAI} + \mathsf{C}^{\mathsf{E}}$. Thus we get the equivalent logic $\mathsf{E} + \mathsf{CC} + \diamondsuit\,\mathsf{ID} + \mathsf{M}^{\mathsf{E}} + \diamondsuit\,\mathsf{CMon} + \mathsf{CA} + \mathsf{CAI} + \mathsf{C}^{\mathsf{E}} + \diamondsuit\,\mathsf{CV}$. By (4) of the same lemma, $\diamondsuit\,\mathsf{CMon}$ is redundant. By (6) we can replace $\diamondsuit\,\mathsf{CV}$ by CV and by (7), we can remove C^{E}. We get: $\mathsf{E} + \mathsf{CC} + \diamondsuit\,\mathsf{ID} + \mathsf{M}^{\mathsf{E}} + \mathsf{CA} + \mathsf{CAI} + \mathsf{CV}$.□

We may conclude as follows: Take any of your preferred vacuist semantics or logic for a basic conditional $>$. One can then always define the neutral conditional $\varphi \gtrdot \psi$, by $(\varphi \gtrdot \psi)^{\circ} = (\varphi^{\circ} > \psi^{\circ}) \wedge \neg(\varphi^{\circ} > \bot)$. Furthermore, with the help of the back-translation $(\varphi > \psi)^{\bullet} = (\varphi^{\bullet} \gtrdot \psi^{\bullet}) \vee \neg(\varphi^{\bullet} \gtrdot \top)$, one can always obtain the neutral analogues of vacuist axiom schemes. Importantly, since we used a very weak basic logic $\mathsf{CM} + \mathsf{CN}$ for $>$, this can be done in a variety of semantics. For example for probabilistic threshold semantics (Hawthorne, 2014), where the neutral conditional $\varphi \gtrdot \psi$ would roughly hold iff $P(\psi|\varphi) \geq t$ and $P(\varphi) > 0$. By this, we can figure out the neutral analogue to the basic probabilistic logic Q.

6 Conclusion

This article proposed a general technique to transfer completeness results of a known conditional $>$ to a definable conditional $\rhd$. The technique was implemented for the *neutral conditional* $\varphi \gtrdot \psi := (\varphi > \psi) \wedge \neg(\varphi > \bot)$. But other definable conditionals can be treated by the same method. For example, Spohn's (2015) *sufficient reason* or Rott's (1986; 2019) *difference making* conditional $\varphi \gg \psi := (\varphi > \psi) \wedge \neg(\neg\varphi > \psi)$, and Spohn's *necessary reason* $\varphi \geq \psi := (\neg\varphi > \neg\psi) \wedge \neg(\varphi > \neg\psi)$ are analyzed in Raidl (2020), the evidential conditional $\varphi \rhd \psi := (\varphi > \psi) \wedge (\neg\psi > \neg\varphi)$ in Raidl, Iacona, and Crupi (2020) and Raidl (2019b). One can also treat Rott's (2019) *dependency conditional* or Lewis's (1973a) *counterfactual dependency* $\varphi \succ \psi := (\varphi > \psi) \wedge (\neg\varphi > \neg\psi)$, and even counterpossible conditionals by a definition of the following sort: $\varphi >' \psi := (\varphi \gtrdot \psi) \vee (\varphi >_G \psi)$. And the list probably continues.

References

Arló-Costa, H., & Shapiro, S. (1992). Maps between conditional logic and non-monotonic logic. In B. Nebel, C. Rich, & W. Swartout (Eds.), *Principles of knowledge representation and reasoning: Proceedings of the third international conference* (pp. 553–565). San Mateo, CA.: Morgan Kaufmann.

Benferhat, S., Dubois, D., & Prade, H. (1997). Nonmonotonic reasoning, conditional objects and possibility theory. *Artifcial Intelligence*, *92*, 259–276.

Chellas, B. F. (1975). Basic conditional logic. *JPL*, *4*(2), 133–153.

Crupi, V., & Iacona, A. (2019). The evidential conditional. Retrieved from `http://philsci-archive.pitt.edu/16479/`

Dubois, D., & Prade, H. (1994). Conditional objects as nonmonotonic consequence relations. In J. Doyle, E. Sandewall, & P. Torasso (Eds.), *Principles of Knowledge Representation and Reasoning* (pp. 170–177). Morgan Kaufmann.

Hawthorne, J. (2014). New Horn rules for probabilistic consequence: Is O+ enough? In S. O. Hansson (Ed.), *David Makinson on classical methods for non-classical problems* (pp. 157–166). Dordrecht: Springer.

Kraus, S., Lehmann, D. J., & Magidor, M. (1990). Nonmonotonic reasoning, preferential models and cumulative logics. *Artificial Intelligence*, *44*(1), 167–207.

Lewis, D. (1971). Completeness and decidability of three logics of counterfactual conditionals. *Theoria*, *37*(1), 74–85.

Lewis, D. (1973a). Causation. *Journal of Philosophy*, *70*(17), 556–567.

Lewis, D. (1973b). *Counterfactuals*. Oxford: Blackwell.

Raidl, E. (2019a). Completeness for counter-doxa conditionals – using ranking semantics. *The Review of Symbolic Logic*, *12*(4), 861–891.

Raidl, E. (2019b). Quick Completeness for the Evidential Conditional. Retrieved from `http://philsci-archive.pitt.edu/16664/`

Raidl, E. (2020). Definable Conditionals. *Topoi,* Special issue*: Foundational Issues in philosophical semantics*. doi: https://doi.org/10.1007/s11245-020-09704-3

Raidl, E., Iacona, A., & Crupi, V. (2020). The Logic of the Evidential Conditional. *Manuscript*.

Rott, H. (1986). Ifs, though and because. *Erkenntnis*, *25*(3), 345–370.

Rott, H. (2019). Difference-making conditionals and the Relevant Ramsey Test. *The Review of Symbolic Logic*, 1–39.

Spohn, W. (2015). Conditionals: A unifying ranking-theoretic perspective. *Philosophers' Imprint*, *15*(1), 1–30.

Stalnaker, R. C. (1968). A theory of conditionals. In N. Rescher (Ed.), *Studies in Logical Theory (American Philosophical Quarterly Monographs 2)* (pp. 98–112). Oxford: Blackwell.

Williamson, T. (2007). *The philosophy of philosophy*. Oxford: OUP.

Intentionality as Disposition*

Yiyan Wang

Department of Philosophy, Tsinghua University, Beijing, China
wang-yy19@mails.tsinghua.edu.cn

Abstract. Regarding collective agency, there have been two conflicting accounts, reducible and irreducible ones. This paper points out that the reason behind these two accounts and their irreconcilable status quo is that there is a tendency towards individualism and even naturalism with exisiting theories. This tendency comes with an unnecessary presupposition, which forces philosophers to face the tension between the irreducible group-concept and ontological monism. We propose a new perspective, that of a dispositional account, which can be considered as an alternative for the individualistic interpretation. We argue that it provides a better solution to the theoretical conundrums.

Keywords: Intentionality · Collective intentionality · Disposition.

1 Introduction

Intentionality itself is a philosophical concept, first introduced by Franz Brentano in the late 19th century:

> *Every mental phenomenon is characterized by what the Scholastics of the Middle Ages called the intentional (or mental) inexistence of an object, and what we might call, though not wholly unambiguously, reference to a content, direction toward an object (which is not to be understood here as meaning a thing), or immanent objectivity.* (Brentano 1874, p.88)

Later, philosophers used intentionality to refer to the aboutness of human consciousness. For instance, Joel believes that it is raining. Joel intends to go to the movies. Joel prefers beer to wine. In each case, Joel is in an intentional state, i.e., one that is about something, refers to something. Intentionality also applies to groups. For example, we intend to go to a bar for a drink. We believe that the study of philosophy is beneficial to mankind. We prefer argumentation to fantasy. In such cases, intentionality shows itself in the first-person plural form and concerns a group's states that are about or refer to something. It appears to be common sense that it is the individual that is the subject of intentionality. Then the core issue is, how do we explain collective intentionality? Can it be structurally reduced to every individual in the group having such intentionality?

* This chapter is in its final form and it is not submitted to publication anywhere else.

B. Liao and Y. N. Wáng (eds.), *Context, Conflict and Reasoning*,
Logic in Asia: Studia Logica Library,
https://doi.org/10.1007/978-981-15-7134-3_12

In other words, if each individual in the group has such intentionality, and these intentions are interdependent, can we infer collective intentionality?

In most existing theories, an intentional state consists of two components: the type of a state and its contents, i.e., psychological modes and propositional contents.[1] Also, intentionality has a subject, namely an entity who possesses the intention. Based on these conceptual features of intentionality, there are three primary types of explanations for collective intentionality: content, mode, and subject accounts.[2] At the same time, we can also classify them as reducible and irreducible accounts according to whether a theory supports the reducible *we-concept*. Approaches that have been influential in this field are Bratman 2014, Gilbert 2006, Searle 2010, Tuomela 2010, List & Pettit 2011.

However, there is a crucial but often overlooked issue: current analyses of collective intentionality are characterized by the interpretation of intentionality as a natural phenomenon. Moreover, it is assumed that this natural phenomenon can be understood by means of the natural sciences (physics, biology, ...). With this explanation, intentionality becomes ontologically homogeneous, and collective intentionality is characterized by individualism, that is, collective intentionality itself may be an independent concept, but intentionality and the ability to intend are individual. Such an ability can only belong to individuals. Thus, in describing intentionality, there is no need to refer to anything beyond the individual itself.

At this point, we may ask whether this tendency toward individualism and naturalism can be justified? if so, how? With the constraints of individualism, can we adequately explain the relationship between individual and collective intention? In this paper, we want to argue that the individualistic explanation cannot sufficiently explain the concept of collective intentionality, and that a possible alternative, a dispositional account, does a better job than the former.[3]

In order to make good on this claim, we need to complete a few steps. First, we need to show that these individualistic claims are indeed made in current analysis. Section 2 will review the most prominent theories in the field and evaluate their solution to the core issue. Second, we need to explain the origins of these individualistic tendencies, and indicate how they lead to adverse outcomes. This will be shown in section 3. Furthermore, in Section 4, we propose an alternative perspective: a dispositional account and argue for its advantages.

[1] For instance, Joel can believe that it is raining, fear that it is raining, or desire that it is raining. In each of these cases, Joel's state has the same propositional content, viz., that it is raining. However, Joel's states are of different intentional types, that is, different psychological modes: belief, fear, desire. etc.

[2] For more details, see the Stanford Encyclopedia of Philosophy article on 'Collective Intentionality' (Schweikard & Schmid 2013)

[3] What we call the dispositional account here, is different from the one in articles about phenomenal intentionality, e.g., Bourget 2010 and Kriegel 2011. The disposition here is an intuitive explanation of intentionality without any presuppositions. The detail will be shown later. Although our idea of disposition comes from quantum physics, we still use the same word, "disposition", as the name of our new non-individualistic explanation, even if it might lead to some ambiguity.

Moreover, we also show that such accounts have had a pervasive influence in other related fields. Let us start with the evaluation of existing theories.

2 An evaluation of the different approaches

In what follows, to identify the problem we evaluate the representative views by prominent philosophers for both the reducible and irreducible accounts. Furthermore, we give a counterexample to demonstrate that neither reductionist nor irreductionist claims can avoid the persistent problem we have seen above.

For the irreducible-account, we introduce John Searle. He takes the core of collective intentionality as "*we-intention*", insists that this notion is irreducible, and gives an argumentation with the well-known business school example:

> *BUSINESS SCHOOL CASE 1*
> *Imagine a group of Business School graduates who were taught and come to believe Adam Smith's theory. After graduation day, each goes out in the world to try to benefit humanity by being as selfish as each of them possibly can and by trying to become as individually rich as they can. Each does this in the mutual knowledge that the others are doing it. Thus there is a goal that each has, and each knows that all the others know that each has it and that they know that each knows that each has it. All the same, there is no cooperation. There is even an ideology that there should be no cooperation. This is a case where the people have an end, and people have common knowledge that other people have that end.*
>
> *BUSINESS SCHOOL CASE 2*
> *All the conditions are the same as in CASE 1, except they together make a solemn pact that they will each go out and try to help humanity by becoming as rich as they can and by acting as selfishly as they can. All of this will be done in order to help humanity.*(Searle 2010, p.47)

Searle argues that CASE 2 is a case of collective intentionality while CASE 1 is not. The reason is that there is an obligation assumed by each individual member in CASE 2, but no such pact or promise exists in CASE 1. Only if such shared obligation exists, can we regard it as a case of collective intentionality. But this kind of cooperation is not *implied* by common knowledge or belief together with individual intentions.

On the other hand, Searle puts this "*we-intention*" only in the mind of an individual. He said explicitly, "The only intentionality that can exist is in the heads of individuals. There is no collective intentionality beyond what is in the head of each member of the collective." (Searle 2010, p.55). In other words, Searle insists on conceptual irreducibility while rejecting ontological irreducible-account. Based on his irreducible "*we-intention*," Searle develops his social ontology, in which people with a certain social status can create social entities through declarative behavior. Obviously, we can see that by being engaged in such kind of declarative behavior, social entities like groups can be created. An everyday example is an enterprise as a legal entity. Thus, Searle's theory clearly has a considerable gap: it allows for collective agency of social entities created by people, but such collective agency cannot instantiate collective intentionality. This diagnosis is also endorsed by (Baier 1997; Stoutland 1997; Meijers 2003).

We agree with Searle's argument that *we-intentions* are irreducible, and we also agree that declarative behavior can produce social entities. However, we disagree with Searle's assertion that the irreducible *we-intention* can only exist in the individuals. This is a clear instance of Searle's commitment to individualism. Nevertheless, Searle's theory does make important contributions. It provides a strong argument for the irreducibility of "*we-intention*," and it outlines a basic frame for a social ontology. However, although the irreducible-account represented by Searle supports conceptually irreducible collective terms, it still presupposes an individualist view on their origins. Let us look at the other side.

In favour of a reducible-account, Michael Bratman argues that there is no need to introduce novel conceptual, metaphysical, or normative notions to explain the relation between joint actions and an individual's single action. He limits his attention to "modest sociality", namely, small-scale cases of shared agency by groups of adults that remain constant over time that do not have authority relations to one another. He claims that his theory of shared agency can reduce Searle's "*we-intention*" by providing sufficient reductive conditions:

A. *Intention condition: We each have intentions that we J; and we each intend that we J by way of each of our intentions that we J and by way of relevant mutual responsiveness in sub-plan and action, and so by way of sub-plans that mesh.*
B. *Belief condition: We each believe that if the intentions of each in favor of our J-ing persist, we will J by way of those intentions and relevant mutual responsiveness in sub-plan and action; and we each believe that there is interdependence in persistence of those intentions of each in favor of our J-ing.*
C. *Interdependence condition: There is interdependence in persistence of the intentions of each in favor of our J-ing.*
D. *Common knowledge condition: It is common knowledge that A-D.*
E. *Mutual responsiveness condition: Our shared intention to J leads to our J-ing by way of public mutual responsiveness in sub-intention and action ...*(Bratman 2014, p.103)

We can see that Bratman's approach gives a structural account of reducibility, one that is clearly individualistic.[4] We agree with (Schmid 2005; Petersson 2007), who argue that Bratman's explanation faces a circularity: an individual

[4] By "structural account" we mean any approach in which intentionality is explained in terms of structural properties. This kind of explanation is familiar from the sciences and naturally related to individualism and even naturalism. However, there is no evidence that intentionality has structural properties by itself. It is more of a hypothesis than a factual explanation. For more details, see "hypothetico-structural explanation" in McMullin 1978.

cannot refer to a joint activity without that joint activity existing, which means that such references cannot cause that joint activity to exist. Beyond that, we should also take a look at the problems in Bratman's reductive conditions. Although Bratman claims that he has given sufficient conditions to characterise collective intentionality, his theory is not enough to get rid of Searle's irreducible "*we-intention*." We will show it by means of a counterexample. Let us go back to Searle's business school example, but modify it a bit so that it can qualify for Bratman's small-scale condition. Instead of leaving school, all students go to the same room after graduation and make online investments to achieve their default goals. In this case, the scale is small enough in Bratman's sense, where each individual is equal, and students do not form companies or corporate organizations. At the same time, this example still preserves Searle's distinction, viz., there is an obligation assumed by each individual member in modified CASE 2, but not in modified CASE 1.

Furthermore, by meeting Bratman's condition, and by the construction of the business school with the modified same room restriction, we can immediately see that the students do have the intention that everyone is striving toward the same goal, and that they do have mutual knowledge that everyone else is striving toward the same goal. The students have a mutual sense of what each other's subplans are. (They are now in the same room). Each student really believes that everyone is pursuing the same goal and that all students are responsive to each other's plans and behavior, and they all believe that their intentions to stick to their goals are interdependent. Moreover, there is indeed an interdependence between their intentions to stick to their goals, and all of the above is common knowledge among them. Indeed, students in the same room react to each other's sub-intentions, and the purpose of the shared action can be traced from individual intentions (Both the group and each individual, in the long run, can achieve the desired goal).

This shows that the modified CASE 1 satisfies Bratman's conditions A-E; the only thing we should look into is the object of intention: Bratman's *we-J*. The question is, what is the meaning of *we-J*? Can we get it from the relationship between individuals?

Bratman uses the form "I intend that *we-J*", where J refers to a joint activity in which the intending member participates. We can check an earlier version of Bratman's concept of intentionality:

We intend to J if and only if:

1. *(a) I intend that we J and (b) you intend that we J.*
2. *I intend that we-J in accordance with and because of 1(a), 1(b), and meshing subplans of 1(a) and 1(b); you intend that we J in accordance with and because of 1(a), 1(b), and meshing subplans of 1(a) and 1(b).*
3. *1 and 2 are common knowledge between us.*(Bratman 1999, p.121)

Thus, we intend J only if each of us separately intends *we-J* and it is common knowledge that each of us separately intends *we-J*. What changed here is just the subject of intentionality; the object of intentionality is still the same, *we-J*. Can this joint activity *we-J* be reduced? Whether the answer is affirmative or negative, the fact remains that Bratman's conditions are not adequate:

- If Bratman holds a radical individualistic view and gives a thorough structural explanation, then he would answer that *we-J* can be simplified. Then, the only unclear concept in our analysis of the example, *we-J*, is also reduced into the individual level; This *we-J* is nothing more than the intention of each student regarding their goal. Consequently, Bratman's condition fails in the test of our modified example, because we still fail to generate a group intention from individual intentions (in Searle's sense).
- If Bratman answers that we-J is irreducible, then it seems that the modified example does produce collective intentionality. However, if we jump out of the example to the theory itself, we see that Bratman has acknowledged an irreducible concept, and his standpoints become indistinguishable from that of Searle. This interpretation just translates Searle's irreducible "we-intention" into irreducible "we-J."

Therefore, we believe that Bratman's analysis is insufficient to refute Searle's argument that there is an irreducible "*we-intention*." That is to say, irreducibility does exist in this field, and with it, an irreconcilable tension between individualism and irreducibility arises. In summary, we claim that Bratman's theory has a strong tendency towards individualism, and that his conditions for collective intention are insufficient. Nevertheless, Bratman's theory has made outstanding contributions, especially regarding the power of the mutual connection among individual intentions.

To sum up, we can see that individualism has a profound influence in this field. Both reducible and irreducible accounts insist on an individualistic view to explain collective intentionality, and both face the intractable tension between the irreducible concept and the reducible ontology. So, why would one insist on individualism, does it bring any beneficial results to the explanation of collective intentionality? In view of the discussion above, there does not seem to be any. Rather, it is precisely because of the presupposition of individualism that all explanations have to face the difficulty of circularity. As we claimed at the beginning, the individualistic explanation cannot adequately explain the concept of collective intentionality, and we believe a dispositional account would do a better job. Before we get to our new account, let us spend some time to look into the reasons behind the widespread individualistic tendency in the field.

3 Naturalism and representation

The term "naturalism" stems from early twentieth century American philosophy, including John Dewey, Ernest Nagel, etc. They claimed that reality is entirely explained by nature, without anything "supernatural," and they advocated scientific methods to investigate every aspect of reality (for more history, see Kim

2003). This paper uses the word "naturalism" in that sense. Naturalism has one premise and two main characteristics. The one premise is this: the natural world is physical. Two characteristics derive from this premise, namely methodological monism, and ontological monism. Methodological monism seeks to reduce explanation to just one type. Ontological monism emphasizes that everything is made up of the same material element.

When we ask, why do people have intentions?, the intuitive answer could be that people can think, and that their thinking reflects the outside world, refers to the outside world, and, more generally, is a representation of the outside world. Then one may ask, why do people think? The intuitive answer could be that people have minds and that this is one of the inherent attributes of human beings. And then one may ask, why do people have a mind? The intuitive answer could be that we human beings have brains. Furthermore, such brains, from a biological point of view, provide us with the material basis for the thinking abilities that make human consciousness possible. This is probably the reasoning process of the individualistic explanation.

We claim that the reason behind the individualism held by the philosophers in this field is that, to some extent, they all stick to the idea of intentionality as representation.[5] Moreover, the underlying motivation behind this view and all of the intuitive answers above are rooted in a naturalistic explanation.

There are several arguments in support of our claim: First, the representational interpretation corresponds to an assumed duality of the mind-world. As a result, minds become the bearer of intentionality as representation. However, when it further leads to the assumption that only a conscious human individual (a natural entity) can act as the bearer of intentionality, we see that this follows the basic naturalistic view that there must be a natural basis for these minds; after all, minds cannot be "supernatural". Second, these representational answers tend to look for a unifying natural foundation on which representation as a non-entity can rest. In fact, this is in accordance with the second characteristic, that of ontological monism, according to which only the physical brain can be the subject that possess intentionality as representation. Third, all of these explanations and answers tend to seek a unified explanation that covers all of the questions, which is actually the first characteristic, that of methodological monism. Combined with the description of the individualistic tendency in Section 1, it is not difficult to see that in the analysis of collective intentionality, naturalism and individualism have the same appeal. The former seeks a natural basis for the "supernatural" intention, while the latter claims that only individuals can have intentionality. These two underlying tendencies together contribute to the representational interpretation and naturally tend to presuppose a structural interpretation.

[5] It should be noted that the close link between intentionality and representationalism is more typical of the analytic tradition. We can find this tendency already in, e.g., the work of Searle, Ryle, Anscombe, and other earlier writers. It is a big topic, and we will continue to look at it more in future papers in this direction.

Based on the chain of representationalism-individualism-naturalism, the subject of intentionality can only be an individual because no one has ever seen a collective as an entity in the real world. Such collective entities have no brains, thus cannot be conscious. According to naturalism, they cannot have representations of the external world, nor can they consciously generate behavioral motives, much less do they have a coherent rationality of their own. In such a vision, there is no collective agency, let alone the collective subject of collective intentionality. Therefore, when it comes to things of different types that do share a common material basis, as in the field of collective agency, naturalism will prioritize reducible accounts as a solution.

It is not hard to see that, individualism is reflected in the interpretation of intentionality as representation, and that it is naturalism that presupposes a unitary material world. Under the guidance of such an ideology, it is not very easy to explain conceptual irreducibility appropriately. Therefore, in the next section, we try to come up with an alternative way, viz., a dispositional explanation, which can avoid such a problem.

4 An alternative: disposition

Based on the above discussion, we can see that both reducible and irreducible theories have a tendency towards individualism and run into severe problems. Therefore, we need to find a system that is consistent with irreducible "*we-intention*" while casting the individualistic tendency aside. In addition, it is necessary to note that even though "*we-intention*" is irreducible, it has a close relationship with individuals.[6]

The traditional individualistic explanation is based on natural science and sticks to the view that all concepts can be explained in a structural way. However, from the current scientific perspective, the basic concepts of natural science are not always structural. For example, in quantum theory there are irreducible dispositions that underlie the microworld. And we can use this irreducible notion to establish our resistance to the traditional individualistic tendency.

From now on, we no longer take intentionality as representational, but as something dispositional. That is, an agent intends p if and only if the agent has a disposition to act in such a way as to bring about p. The relevant dispositions (of both individuals and groups) here are the disposition to act (pro-actively or re-actively, as the case may be). As the expression of a certain tendency, the disposition only manifests itself in time-space and does not require us to commit ourselves to any prior conditional hypothesis to ensure its rationality. Intentionality as disposition does not correspond to a hypothetical mind-world duality, and the disposition itself does not necessitate reliance on a material basis. From the perspective of a dispositional-account, the intentional subject can be more appropriately explained. When we ask about the subject of an intention, there is no reason to go back to naturalism in which we have to have a brain in

[6] Admitting the irreducibility of "*we-intention*" and rejecting an ontological reducible-account will make the whole system look more like an emergentism theory.

the natural sense to become an agent. In this new interpretation, intentionality, as a disposition, can be owned by individuals or groups.

We believe that through this explanation, we can build a unified model to eliminate the tension and inconsistency in the existing systems. Consider the earlier business school example, if we explicitly interpret intentionality as something dispositional, we will accept an ontological pluralism and acknowledge that in CASE 1, the subject of each intention is individual, while in CASE 2, the "*we-intention*" subordinates to the whole business school in parallel with each individual intention that belongs to the individuals. Although this "*we-intention*" has many connections with the individual intentions which can analysed in terms of conditions listed by Bratman, it is still not a reducible concept. And with this interpretation, we do not need to go to the pains of explaining how the collective intention only exists in individual consciousness like Searle has to do, since there is nothing presupposed about the location of intentionality. As we can see, the dispositional account does address the core issue, and because it does not have redundant assumptions concerning a physical basis, it does so well. Note that group intentionality is inextricably related to individual intentionality. Then, of course, no one can stop people from saying that collective dispositions are partly based on individual dispositions. However, as we saw in the debate between Searle and Bratman in Section 2, we cannot obtain collective intentionality completely through individuals; that is, the concept of collective intentionality is irreducible. Therefore, we cannot completely reduce the disposition of a collective to that of its members' in our dispositional account.

Several advantages can be gained by adopting the dispositional account: First, the dispositional account makes intentionality break away from the duality of the mind-world. As a neutral explanation, the dispositional account is neither extreme individualism nor extreme collectivism. It no longer presupposes that intentionality is an attribute that belongs only to individuals who have a physical container of consciousness, viz., a brain. Second, on a conceptual level, collective intention has the same ontological status as individual intention, and there is no question of which constitutes or decides which. Third, the original structural explanation of collective intentionality has a kind of mechanical sense and does not show the directedness of intentionality, whereas the dispositional account can do that more intuitively and naturally. All in all, within the context of collective intentionality, it seems much more intuitive to look at it in terms of disposition rather than structure, and if we do so, we can incorporate it into a rich and varied social ontology consistently.[7]

At this moment, we may take one step back. For reasons of space we have not had the opportunity to dig into the hardest problem, which is the ontological,

---

[7] Note that our definition of disposition is different from the one in the field of phenomenal intentionality, see the Stanford Encyclopedia of Philosophy article on 'Phenomenal Intentionality' (Bourget & Mendelovici 2017). We have noticed some attacks on the dispositional account, such as (Schiller 2019). However, for reasons that we cannot present here due to space restrictions, we are confident that such an attack will not work against our conception.

metaphysical issue. Therefore, we should remain cautious about what we don't know. The purpose of this paper is not to dismantle a premise and install a new one, but to show that in this area, there are things that people take for granted but that are actually uncertain. As a step back, we want to demonstrate that similar research methods have already been employed in other fields, such as linguistics and formal epistemic logic, to support our standpoint. For linguistics, (Stokhof et al. 2016) trace some assumptions that have informed conservative naturalism in linguistic theory since Chomsky. The paper argued that behind naturalism lies the imitation of natural science, and that language, which is not only material stuff, can be explained in a dispositional way, in which linguistic research can expand to a broader field. In formal epistemic logic, (Baltag et al. 2019) show how the group's belief state tends to form, which is determined by the group's trust matrix and agents' initial belief state. Furthermore, through a formal epistemic logical system, they claimed that the notion of potential group belief hinges on the tendency of the group. Which can be counted as technical support for the dispositional interpretation.

5 Conclusion and ideas for a future formal investigation

Once a dispositional account is adopted, then at least in the topic of collective agency, we must admit that collective entities are ontologically heterogeneous and cannot be explained uniformly. This will eventually lead to an ontological pluralism, in which different categories of natural entities cannot be reduced to each other. This future prospect also fundamentally supports our view: "*we-intention*" is ontologically different from individual intention; any attempt to reduce one to the other is infeasible.

We do not assume that structural explanations are useless, nor that the dispositional account is the final answer. This is an open topic. In the future, we will continue to carry out more formal theoretical research, to reconcile the tensions, to build a unified, consistent theory. Many other formalized research directions can be continued, too, such as the rationality of collective agency, the connection with formal semantics, and the formal system based on it.

We must admit that we have not been able to discuss the complex issues of the ontological existence of collective entities so far. Therefore, this article only wants to put forward that, when the individualist analysis is confronted with a conceptual irreducible tension, we offer an alternative way, namely, a dispositional account, that can provide a better explanation.

Acknowledgement This research is supported by the Major Program of the National Social Science Foundation of China (NO. 17ZDA026). I want to thank Fenrong Liu and Martin Stokhof for their extensive comments, and discussions with me on the earlier versions of the paper. I want to thank the two AWPL anonymous referees for their helpful comments.

References

1. Baier, A.: Doing Things With Others: The Mental Commons. In L. Alanen, S. Heinämaa, and T. Wallgren (eds.), Commonality and Particularity in Ethics, pp. 15–44. St. Martin's Press, New York (1997).
2. Baltag, A., Liu, F., Shi, C., Smets, S.: Collective Belief as Tendency Toward Consensus. Manuscript (2019).
3. Bourget, D.: Consciousness is underived intentionality, Noûs, 44(1): 32–58 (2010).
4. Bourget, D., Mendelovici, A.: Phenomenal intentionality. The stanford encyclopedia of philosophy (2019).
5. Bratman, M.: Faces of Intention: Selected Essays on Intention and Agency. Cambridge University Press, Cambridge (1999).
6. Bratman, M.: Shared Agency: A Planning Theory of Acting Together. Oxford University Press, Oxford (2014).
7. Bratman, M.: Planning, Time, and Self-Governance: Essays in Practical Rationality. Oxford University Press, Oxford (2018).
8. Brentano, F.: Psychology from an Empirical Standpoint (Psychologie vom empirischen Standpunkt). Routledge and Kegan Paul, London (1874).
9. Gilbert, M.: A Theory of Political Obligation: Membership, Commitment, and the Bonds of Society. Oxford University Press, Oxford (2006).
10. Kim, J.: The American Origins of Philosophical Naturalism. Journal of Philosophical Research, APA Centennial Volume: 83–98 (2003).
11. Kriegel, U.: The sources of intentionality. Oxford University Press, Oxford (2011).
12. List, C., Pettit, P.: Group Agency: The Possibility, Design, and Status of Corporate Agents. Oxford University Press, Oxford (2011).
13. McMullin, E.: Structural explanation. American Philosophical Quarterly. 15 (2), 139-47. University of Illinois Press, Carbondale (1978).
14. Meijers, A.: Can Collective Intentionality be Individualized? In: American Journal of Economics and Sociology 62: 167–83 (2003).
15. Petersson, B.: Collectivity and Circularity, The Journal of Philosophy, 104(3): 138–56 (2007).
16. Schiller, H.I.: Phenomenal dispositions. forthcoming in Synthese, 1–12. http://dx.doi.org/10.1007/s11229-018-01909-9.
17. Schmid, H.B.: Wir-Intentionalität. Kritik des ontologischen Individualismus und Rekonstruktion der Gemeinschaft. Freiburg i.Br., Alber (2005).
18. Schweikard, D., Schmid, H.B.: Collective Intentionality. The stanford encyclopedia of philosophy (2013).
19. Searle, J.: Intentionality: An Essay in the Philosophy of Mind. Cambridge University Press, Cambridge (1983).
20. Searle, J.: Collective Intentions and Actions. in P. Cohen, J. Morgan, and M.E. Pollack (eds.): Intentions in Communications. Cambridge, MIT Press, Mass. (1990).
21. Searle, J.: Making the Social World: The Structure of Human Civilization. Oxford University Press, Oxford (2010).
22. Stokhof, M., van Lambalgen, M.: What Cost Naturalism? In: K. Balogh, W. Petersen (eds), Bridging Formal and Conceptual Semantics. Selected papers of BRIDGE-14, pp. 89–117. Düsseldorf University Press, Düsseldorf (2016).
23. Stoutland, F.: Why are Philosophers of Action so Anti-Social? In L. Alanen, S. Heinämaa, and T. Wallgren (eds.), Commonality and Particularity in Ethics, pp. 45–74. St. Martin's Press, New York (1997).
24. Tuomela, R.: The Philosophy of Sociality: The Shared Point of View. Oxford University Press, Oxford (2010).

The manufacturer's authorised representative in the EU is Springer Nature Customer Service Centre GmbH, Europaplatz 3, 69115 Heidelberg, Germany. If you have any concerns regarding our products, please contact ProductSafety@springernature.com

Printed and bound by CPI Group (UK) Ltd, Croydon, CR0 4YY
30/09/2026
02225513-0016